ENLIGHTENED
ENVIRONMENTALISM

ENLIGHTENED ENVIRONMENTALISM

HOW WE GOT HERE AND
HOW TO RESCUE OUR FUTURE

ANAND M. SAXENA, Ph.D.

Universal-Publishers
Irvine • Boca Raton

Enlightened Environmentalism:
How We Got Here and How to Rescue Our Future

For permission to photocopy or use material electronically from this work, please access www.copyright.com or contact the Copyright Clearance Center, Inc. (CCC) at 978-750-8400. CCC is a not-for-profit organization that provides licenses and registration for a variety of users. For organizations that have been granted a photocopy license by the CCC, a separate system of payments has been arranged.

Universal Publishers, Inc.
Irvine • Boca Raton
USA • 2021
www.Universal-Publishers.com

ISBN: 978-1-62734-356-5 (pbk.)
ISBN: 978-1-62734-357-2 (ebk.)

Typeset by Medlar Publishing Solutions Pvt Ltd, India
Cover design by Ivan Popov

Library of Congress Cataloging-in-Publication Data

Names: Saxena, Anand M., author.
Title: Enlightened environmentalism : How we got here and
    how to rescue our future / Anand M. Saxena.
Description: Irvine : Universal Publishers, [2021] | Includes
    bibliographical references.
Identifiers: LCCN 2021028946 (print) | LCCN 2021028947 (ebook) | ISBN
    9781627343565 (paperback) | ISBN 9781627343572 (ebook)
Subjects: LCSH: Environmental degradation. | Conservation of natural
    resources. | Environmentalism.
Classification: LCC GE140 .S294 2021 (print) | LCC GE140 (ebook) | DDC
    304.2/8--dc23
LC record available at https://lccn.loc.gov/2021028946
LC ebook record available at https://lccn.loc.gov/2021028947

I gratefully acknowledge the help given by my daughter Nalini in all phases of the development of my ideas into this book.

CONTENTS

INTRODUCTION

Technological progress and globalization have changed our lives by providing us with numerous comforts and conveniences. While factories keep churning out new gizmos at regular intervals that dazzle and delight us, we get objects of daily use, including clothing, furnishings, edible items, even fruits and flowers, from distant regions of the world. At the behest of industries, scientists and technologists keep creating objects with fascinating and amazing features. However, there are clouds on the horizon that clearly indicate that this era of profligate consumption has caused enormous harms to the ecosystem that may threaten our welfare, perhaps even survival.

Fossil fuels and their derivatives changed the lives of people almost everywhere. Personal transportation increased the range of human habitation, animal factories increased the production and consumption of meats and dairy, agriculture became highly dependent on chemical products, and electricity, mainly produced from fossil fuels, became an essential part of life. At the same time, products made with fossil fuels—from plastics to pharmaceuticals—filled our homes. The realization came later that a price had to be paid for these conveniences in terms of climate change and its deleterious effects that endanger human life and welfare. Events such as droughts, irregular rainfalls, forest fires, and intense hurricanes cause substantial damages, including loss of lives. Also, there are changes that are building up and will have serious consequences with the passage of time. Melting of glaciers and polar ice caps will accelerate and will have disastrous consequences in many parts of the world. Increasing acidity of oceans due to dissolved carbon dioxide from the atmosphere will have an adverse effect on marine life everywhere.

There are many other clouds on the horizon that threaten the welfare of humanity. We have lived with the implicit belief that the Earth is too large to be affected by human actions and has an unlimited capacity to provide us with resources and absorb the waste produced by our activities. With this mindset, oceans were freely used for dumping undesirable items, effluents from factories were only considered to be a temporary and local nuisance, and forests were razed whenever a more profitable use of the land was discovered. Agriculture was supposed to provide food items forever, albeit with some help from synthetic fertilizers and other chemicals. We have now come to realize that there are consequences of this lack of concern for the biosphere. Air and water pollution kill or shorten the lifespan of millions of people in all parts of the world, including the United States. Degradation of farmlands due to the loss of topsoil and the accumulation of chemicals is decreasing their productivity, while the population is increasing at a steady pace and the demand of agricultural products is increasing both for direct consumption by humans and also to feed the huge number of farm animals. A byproduct of our profligate lifestyle is that we are producing waste on a gargantuan scale that cannot be assimilated by the ecosystem. Ignorance of planetary limits and blind faith in technology are projected to adversely affect our welfare within the next few decades. Scientists and ecologists began to realize some time ago that this lifestyle is not sustainable, but the engines of industry were moving at full speed and reversing direction, or even slowing their growth, was considered to be suicidal by many of them. Continuously increasing consumption is maintained by inducements from businesses in the form of advertisements, easy credit, and rapidly changing features of consumer products. It is also encouraged by governments because personal consumption constitutes a major portion of the economy.

Environmental degradation and depletion of planetary resources may make life difficult, even impossible, for a large segment of the human population within a few decades unless drastic corrective actions are taken soon. The cost of mitigating steps will rapidly increase if actions are not taken expeditiously. Humans have proved numerous times that they are very resilient and can adapt to extreme changes when they perceive a threat to their existence. For example, the threat of COVID-19 has changed the behavior of people in all parts of the world. There is evidence that consciousness of the threat to human existence is developing now, and some people have started reducing their ecological footprint. The important thing is to increase the awareness of the present situation before it becomes too late to stop the progression of events. Steps taken to ameliorate the development of dangerous events may require reducing consumption to sustainable levels, which will be vigorously opposed by people who control big businesses. Another problem is that many deleterious developments are not connected, in the common perception, with our profligate lifestyle and also there is the misplaced belief that science and technology will solve all problems in the course of time. Although technological developments can provide us with incredible gadgets for everyday use, the problems of decreasing productivity of farmlands, intense hurricanes and thunderstorms, droughts, rising sea levels, and the loss of biodiversity are global problems that need concerted actions by people everywhere.

The essays in this book develop an interrelationship between these developments with suggestions for making changes. The burden of humanity on the planetary ecosystem has increased at a very rapid rate during the last few decades, with the result that we are handing over a deeply flawed planetary system to the younger generation. Many young people in different parts of the world have come to the realization that they will have to face the consequences of the profligate lifestyle of previous generations. Millions of young people all over the world are demanding action from governments to fight global warming and environmental degradation. In September 2019, 4 million marchers in many countries demanded swift action from governments. Greta Thunberg, one of the leaders of the movement, addressed the United Nations General Assembly: "You have stolen my dreams. All you can talk about is money and fairy tales of eternal economic growth." Unfortunately, the epidemic of COVID-19 diverted the resources of people and governments to combating it, and the Green Agenda became of secondary importance. However, the environmental problems that have been building up for decades will not take a break and their effects will become even more calamitous due to lost time. It is up to us to save the planet from extreme events for our sake and for the sake of future generations. As an ancient Native American proverb says: "We do not inherit the Earth from our ancestors, we borrow it from our children."

ONE

·····················

CLIMATE CHANGE

C limate change is the greatest threat facing humanity because it endangers the welfare and lives of people everywhere in the world. Using satellites, weather balloons, floating buoys, and other instruments to monitor changes in temperature, scientists and engineers have carefully collected data over the last few decades in all parts of the world—from the depths of oceans to the top of the atmosphere. Their evidence is unequivocal: our planet is warming and the average global temperature breaks records almost every year. The initial consequences of global warming have already started affecting human lives while other consequences will endanger the lives and livelihoods of millions of people in the coming decades. Although a change in temperature by a degree or two may seem trivial, warming on a global scale will have drastic consequences. The number and severity of the adverse effects of global warming are increasing with each passing year. Extreme events such as floods, hurricanes, storms, forest fires, and droughts have become common in many parts of the world with devastating effects. Rising sea levels have swallowed a few islands and many coastal regions are in danger of being submerged. Droughts and higher temperatures have decreased the output of farms in many places. Increased acidity of oceans, disappearance of glaciers, and melting of ice on snow-covered peaks will cause death and destruction on a large scale.

Fossil Fuels

The main cause of global warming is the concentration of carbon dioxide in the atmosphere. A dynamic equilibrium between the sources and sinks of carbon dioxide maintained its atmospheric level within a narrow range for millennia.

This gas was primarily emitted from respiration by plants and animals, with some contribution from occasional forest fires, and was absorbed by plants during photosynthesis. The use of wood and charcoal by humans added to the burden of this gas in the atmosphere but it was not large enough to significantly disturb this equilibrium. The discovery and extensive use of fossil fuels upended this balance, resulting in far more carbon dioxide being produced than could be assimilated by natural systems.

Fossil fuels played a crucial role in ushering in the Industrial Revolution that began the process of transforming our lives. In the early stages of industrialization, the primary source of energy was coal, which had a much greater concentration of energy than wood. Coal is still widely used for the generation of electricity in power plants in almost all countries. The discovery of petroleum added another dimension to the use of fossil fuels because gasoline could be easily transported and ignited when needed, giving rise to motorized transportation. Fossil fuels—coal, natural gas, and petroleum—are still the major fuels to produce energy, both in the United States and around the world. In addition to providing a cheap and abundant source of energy, petroleum can also be used to manufacture a wide range of chemicals such as for plastics, fertilizers, pharmaceuticals, detergents, and perfumes. As technologists discovered more and more uses of fossil fuels and their derivatives, the public became fascinated with the products of this new source of energy and chemicals. Industries based on fossil fuels have provided us with resources and amenities that previous generations could not have imagined.

Energy consumption is a key differentiating factor between modern and pre-industrialization societies. Energy consumption generally determines the standard of living of a people. An average American consumes three times more energy than a Mexican, and seventeen times more energy than an African on a per capita basis.[1] Although fossil fuels are not the only available source of energy, other sources of energy have not played a significant role until recent times. The contribution of the alternative sources of energy is expected to grow with increasing awareness of the deleterious effects of fossil fuels.

All aspects of human existence today—food, shelter, transportation, furnishings, and a great variety of stuff that provides us with many types of creature comforts and with which we fill our homes—are either made directly from fossil fuels or by machines that operate by using these nonrenewable resources. Fossil fuels were formed from the organic remains of prehistoric plants and animals that have remained buried for millions of years. The Earth holds an enormous amount of these sources of energy. The use of fossil fuels has been steadily increasing for almost a century. The average per capita consumption in the United States is about 2.7 gallons of oil-equivalent per day.[2] On a world-wide basis, humans use about 10 billion tons of oil-equivalent per year.[3]

Until a few decades ago, no serious consideration was given to the environmental effects of fossil fuels because it was believed that the Earth was too big to be affected by human activities. This line of thinking, and the easy availability of energy and chemicals from fossil fuels, prevented any serious considerations of their effects on the environment. When scientists concluded that the use of fossil fuels is damaging the planetary ecosystem, the engines of industry were moving with such speed that reversing direction, or even slowing the pace, would have been considered financial suicidal by many industries. Hence, the warnings from scientists were ignored and, to some extent, are being ignored even today.

Due to increasing concerns about the effect of burning fossil fuels, the Intergovernmental Panel on Climate Change (IPCC) was set up by the United Nations (UN) and the World Meteorological Organization in 1988. IPCC's objective is to assess the scientific, technical, and socioeconomic information relevant to understanding the risk of human-induced climate change, and to make recommendations to stabilize the concentration of greenhouse gases (GHG) at a level that will prevent dangerous interference with the global climate system. IPCC publishes the opinions of leading climate scientists and the consensus of participating governments. These reports assess the causes of climate change and their likely impact on the planet and have been emphasizing the gravity of the situation with pleas for immediate action to prevent extreme events. However, most nations have taken only faltering steps to decrease the emission of GHG. The longer the nations and people delay definitive actions, the more extreme will be the steps necessary to prevent disastrous events. With each passing year, the situation continues to worsen and the window of opportunity to prevent extreme changes in the environment may be about to close. The Fifth Assessment Report of IPCC warned that the world may be fast approaching a tipping point concerning climate change and suggested that the next few years will be crucial for reducing the amount of greenhouse gases.[4]

Since the use of fossil fuels is the main reason for the precarious situation in which we find ourselves, it is useful to understand the processes for extracting, purifying, and combusting them to produce energy. In addition to the accumulation of greenhouse gases, the very process of extracting them from the ground has an adverse effect on the environment, the health of mineworkers, and people who live in the vicinity of mines. The processes involved and dangers associated with the extraction and purification of fossil fuels are discussed in an appendix.

Greenhouse Gases and Their Global Warming Potential

Greenhouse gases trap heat in the atmosphere, resulting in warming of the Earth. The extent to which a gas is effective in global warming depends on a number of factors, including the concentration of the gas in air, the period for which it stays in the atmosphere, and its capacity to absorb the radiant heat emanating from the Earth. Although carbon dioxide is the primary greenhouse gas, methane, nitrous oxide, and fluorinated gases also make significant contributions to global warming. The capacity of a gas to trap heat in the atmosphere is called its Global Warming Potential (GWP). Carbon dioxide is assigned a GWP of one, and the effectiveness other gases is measured in terms of "carbon dioxide equivalent." Since the period for which a gas stays in the atmosphere varies considerably, the GWP of a gas will also depend on the duration for which its effectiveness is being considered. It is common to consider a time frame of 20 years when comparing the effects of greenhouse gases. Methane (CH_4) stays in the atmosphere for about 12 years and has a GWP of 84 to 87 for this period.[5] This means that this gas is about 85 times more effective in its warming potential than carbon dioxide. Nitrous oxide stays in the atmosphere for about 120 years and has a GWP of 280. Several hydrofluorocarbons (HFCs) are also produced during the combustion of fossil fuels. In general, they are good absorbers of thermal radiation, and their lifetimes range from one year to many centuries, hence their GWPs can be very large. As an example, the GWP of CFC-11 (CCl_3F) is 6,730 for a 20- year time horizon. Consideration of the lifetime of greenhouse gases becomes important when the objective is to determine the long-term effect of gases already in the atmosphere.[6] In 2014, the total GHG emission in the United States was 6,870 million tons of carbon dioxide equivalent.[7] Total emission of greenhouse gases in the world is estimated to be 53,526 million metric tons of CO_2 equivalent per year.[8]

While carbon dioxide is produced in copious amounts during the combustion of fossil fuels, particularly in motor vehicles, other greenhouse gases are released in the atmosphere during various uses of petrochemicals. Human activities that release methane in the atmosphere include livestock farming and leakages in the production, transportation, and use of natural gas. Fracking, the process of injecting highly pressurized liquid into rocks with the goal of extracting oil or gas, is increasing in popularity and leads to the release of large amounts of methane. Other sources of methane include landfills, organic waste, biomass burning, and rice agriculture. Some nitrous oxide is usually produced during the combustion of almost all fossil fuels. This gas is also released from the decomposition of synthetic fertilizers containing ammonia or nitrates in the soil. HFCs are used in refrigerators, air conditioners, foams, and aerosol cans. They frequently leak during the manufacture of these products and continue to leak throughout the product's life. Hydrofluorocarbons are produced by human activities mainly in the industrial processes that use fossil fuels; there is no natural source of these gases.

Global Temperature and Carbon Dioxide

Many natural processes release carbon dioxide in the atmosphere; these include soil decomposition, ocean-atmosphere exchange, volcanic eruptions, and respiration from plants and animals. Plants, however, also absorb carbon dioxide in the presence of sunlight and stabilize it in the form of organic matter in their leaves, stems, roots, and other components. On average, plants absorb much more carbon dioxide by photosynthesis than they release during respiration. For thousands of years—until the mass-scale use of fossil fuels—sources and sinks kept the level of carbon dioxide within a narrow range in a dynamic equilibrium.

Scientists and engineers have explored the link between the concentration of carbon dioxide in the atmosphere and global temperature by drilling into ice sheets that cover Antarctica and Greenland, where thousands of years of snow has compressed into thick slabs of ice. These ice layers also trap bubbles from the ancient atmosphere, thus allowing a direct measurement of the level of CO_2 and other gases in the atmosphere in the years when the ice was formed. While the level of CO_2 is measured directly from the bubbles trapped in ice, the temperature at different periods is inferred from the concentrations of isotopes of oxygen. By determining the ratio of the heavy isotope of oxygen (^{18}O) to the light isotope (^{16}O) in marine sediments, ice cores, and fossils, scientists can determine changes in the atmospheric temperature at the time these objects were formed.[9] These observations show a tight correlation between global temperature and the level of carbon dioxide during the last 800,000 years, as shown in Figure 1.

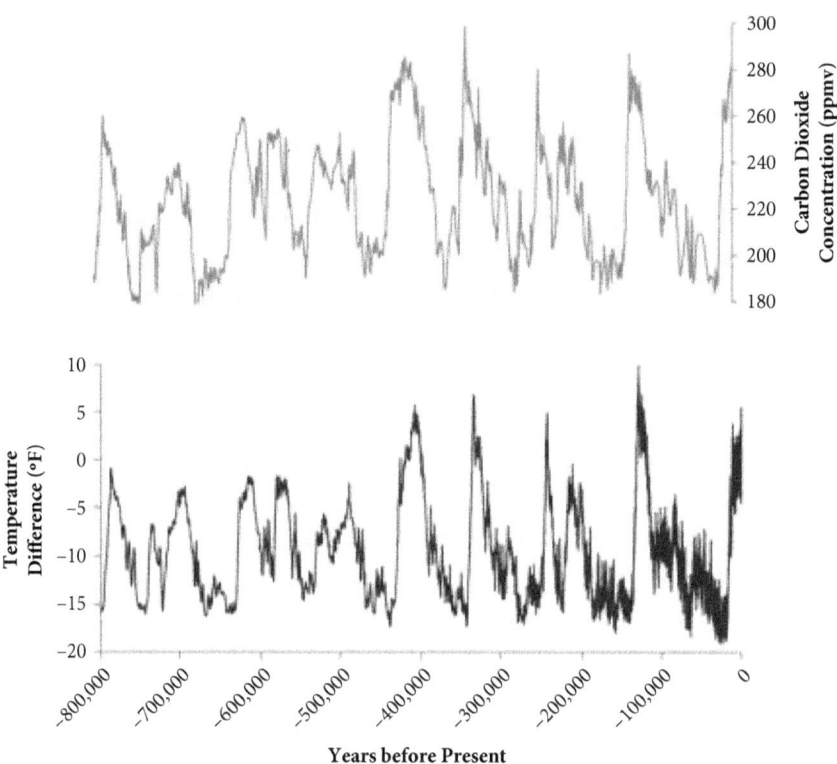

Figure 1 Correlation between CO_2 Level and Temperature over 800,000 years.[10]

The plots show that during the last 800,000 years, there have been seven cycles of significant variation in the amount of carbon dioxide in the atmosphere. During these cycles, the concentration of CO_2 in the atmosphere varied between 180 and 300 parts per million (ppm). This variation is caused by natural events such as volcanic eruptions, ocean-atmospheric exchange, and the formation and dissolution of microscopic forms of life on a global scale.

Since the mid-20th century, humans have been adding huge quantities of CO_2 to the atmosphere by the combustion of fossil fuels. The level of carbon dioxide in the atmosphere is now at unprecedented high levels. Its concentration since 1960 has been determined at the Mauna Loa Observatory on the island of Hawaii. The isolation of this place allows a record of the concentration of this gas that is unperturbed by industrial activities. It shows a continuous increase from 1960 at the rate of 2 ppm per year in recent decades, reaching a level of 410.27 ppm in November 2019. The data obtained from the observatory since 1960 is shown in Figure 2.

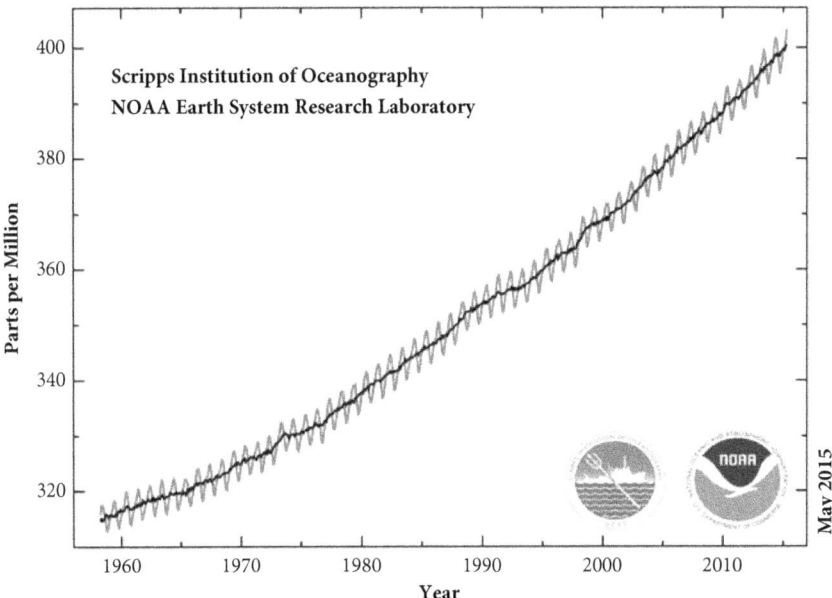

Figure 2 Atmospheric CO_2 at Mauna Loa Observatory.
Source: NOAA.

Figure 3 shows the average global temperature from 1880 to present. The undulations represent the annual mean, and the solid line displays the five-year mean. The average global temperature has been steadily rising since 1970, setting records almost every year. Earth's average temperature has increased by 1.6°F (0.9°C) since 1910 with the major part of that increase occurring during the last three decades. According to the National Aeronautics and Space Administration (NASA) and the National Oceanic and Atmospheric Administration (NOAA), 2016 and 2020 were the warmest years since records were first kept in 1880. The five warmest years have all occurred since 2010.[11]

Since the mid-20th century, the combustion of fossil fuels has released large amounts of CO_2 in the atmosphere. Table 1 shows the amount of carbon dioxide emitted by the combustion of wood and various fossil fuels.[12] The number is highest for wood because it is less dense in energy than fossil fuels. The amount of carbon is not the same in different types of coal, as indicated by the range given in the table. Anthracite coal has the greatest concentration of carbon and is the most efficient producer of energy. Lignite has the least carbon with bituminous coal somewhere in between.

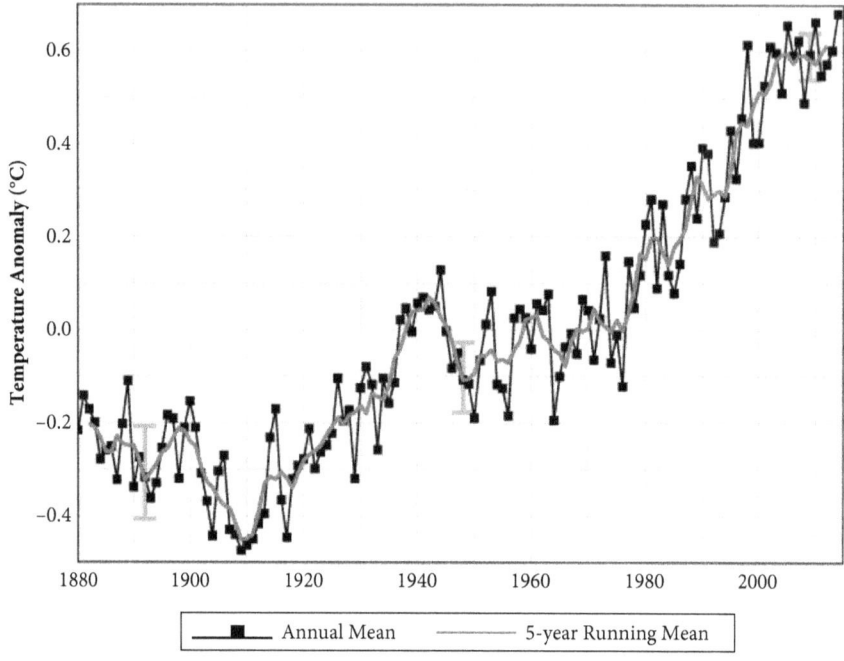

Figure 3 Global Land–Ocean Temperature Index.
Source: NASA.[13]

Table 1 Emission of CO_2.

Fossil Fuel	Emissions in $KgCO_2/GJ$
Wood	109.6
Coal	101.2–94.6*
Fuel Oil	77.4
Diesel	74.1
Crude Oil	73.3
Kerosene	71.5
Gasoline	69.3
Natural Gas	56.1

Note: *depending on the type.

In June 2015, the National Oceanic and Atmospheric Administration released the temperature data obtained from numerous observation stations, buoys, commercial ships, and weather stations. An analysis shows that the temperature of the Earth has been rising continuously for the last fifty years. Although an increase in temperature by a degree or two may be considered trivial because the temperature at most places fluctuates by many degrees during the year, an increase in the average temperature at the global scale will have serious consequences. The close correlation between the level of carbon dioxide in the atmosphere and the global temperature clearly indicates that this warming is related to the level of CO_2 in the atmosphere caused by the combustion of fossil fuels.

Methane and Other Greenhouse Gases

Methane is the second-most-common greenhouse gas. Since it is about eighty-five times more effective than carbon dioxide in trapping the radiant heat emanating from the Earth, it makes a much greater contribution to global warming than its concentration in the atmosphere. The level of methane in the atmosphere has been rising since 2007. Methane leaks at almost every point in the supply chain—at drilling sites, compression stations, and from the network of pipes that deliver natural gas to homes. The complexity of these sources makes it difficult to quantify the amount of methane released in the atmosphere. In 2016, the Environmental Protection Agency (EPA) estimated that the amount of methane released is about 9.3 million metric tons per year, 27 percent higher than the previous estimate. Over a twenty-year timeframe, these emissions have the same impact on the climate as 200 coal-fired power plants. Methane gas leaked from the supply chain is estimated to be worth $1.4 billion at 2015 prices.[14]

Methane is also released in the atmosphere during the hydraulic fracturing process. Shale gas, essentially the same as natural gas, is tightly held in shale formations and is extracted by high-precision horizontal drilling and hydraulic fracturing (fracking). A large amount of methane is emitted during the extraction process, which almost completely negates the advantage of shale gas over coal in its impact on climate change.[15] Methane is also released in the atmosphere in oil and gas operations. Satellite observations of oil and gas basins in Texas and North Dakota show that substantial amounts of methane are leaked into the atmosphere from these facilities. Other sources of methane include decaying organic matter in landfills, rice agriculture, and the digestion of food in the stomachs of ruminants (cows, goats, and sheep). When these animals ingest food, it is fermented in the first division of the stomach—known as the rumen—as part of the digestive process that releases methane. It is estimated that about a quarter of the methane in the atmosphere has its origin in enteric fermentation of food in the stomachs of farm animals. Since humans raise livestock for food, this methane emission is related to the consumption of meat and dairy. When all sources of methane emission are considered, natural gas may be just as damaging for the environment as coal or oil.[16] The EPA estimated that 8 million tons of methane leaks into the atmosphere per year in the U.S. However, the Environmental Defense Fund (EDF) estimated that the leakage rate of methane is much higher at 13 million tons per year.

The atmospheric concentration of greenhouse gases has been increasing steadily since the beginning of the Industrial Revolution, mainly from the use of fossil fuels by individuals, businesses, and industries. The worldwide emission of greenhouse gases increased by 35 percent during the twenty-year period from 1990 to 2010. Carbon dioxide accounted for three-fourths of this total; its concentration in the atmosphere increased by 42 percent during this period. Dr. James Hansen, former head of the Goddard Institute for Space Studies, made the following statement: "If humanity wishes to preserve a planet similar to that on which civilization developed and to which life on Earth is adapted, paleoclimatic evidence and ongoing climate change suggest that CO_2 will need to be reduced from current levels to below 350 ppm."[17] This statement has led to a movement known as 350.org. Since the last recorded level of carbon dioxide was 412.89 ppm in November 2020, the goal of this eminently sensible movement is not only to stop the growth of greenhouse gases in the atmosphere but also to reduce their concentrations to what the levels were in the year 1990.

Scientific Consensus

There is overwhelming scientific consensus that global warming is indeed happening and is caused by the release of greenhouse gases from human activities. In fact, almost the entire concentration of greenhouse gases during the last few decades is attributable to human activities.[18] Eighteen leading American and international scientific societies have joined together to make the following statement, supported by NASA: "Observations throughout the world make it clear that climate change is occurring, and rigorous scientific research demonstrates that the greenhouse gases emitted by human activities are the primary driver."[19] An examination of 11,944 scientific papers on climate change published from 1991 to 2011 showed that 97.1 percent endorsed the consensus position that humans are causing global warming.[20] The few papers that do not support anthropogenic climate change have been shown to have methodological flaws, in that they ignored information that does not fit their conclusions or used inappropriate statistical methods.[21] A more recent analysis of 24,210 peer reviewed papers by 69,406 authors shows that only four authors reject human-caused global warming, which means that the consensus of climate scientist on this topic is 99.99 percent.[22] More than 200 worldwide scientific organizations hold the position that climate change is caused by greenhouse gas emissions from the burning of fossil fuels.

Consequences of Climate Change

Although climate hazards are natural events in the weather cycle, the scale of destruction and devastation caused by extreme climatic events in recent decades is new and terrifying. The frequency of such events is increasing with each passing year, endangering the lives and livelihoods of millions of people in almost all parts of the world. A striking, visible effect of global warming is the melting of ice and snow that cover high mountains. Glaciers, which help maintain the flow of rivers throughout the year, are shrinking in size around the world and most of them are projected to disappear within the next few decades. The areas covered by ice sheets in the Arctic and Greenland have been decreasing each year, and Kilimanjaro, the highest mountain in Africa, has lost 80 percent of its snow cover.

The average global temperature breaks records almost every year. Even a small increase in temperature on the global scale will have many consequences, some of which will affect the lives and livelihoods of millions of people right now, while others will slowly degrade the planetary ecosystem in ways that will have far-reaching effects on the welfare of humanity. Evidence of changes in climate and weather patterns throughout the world has been accumulating for the last few decades. Extreme events, such as high category hurricanes, typhoons, periods of excessive or inopportune rainfalls, forest fires, and droughts have become more frequent with devastating effects on communities around the world. The slow but persistent events that are bound to have profound consequences include rise in sea levels, acidification of oceans, and changes in the flow of rivers caused by irregular rainfalls and the melting of ice caps on mountain peaks.

The list of extreme climatic events that have occurred in the last few years is growing on a continual basis. These events cause economic loss, hardship, and loss of lives in different parts of the world. The frequency and severity of these events is increasing with the passage of time. Data about major disasters in each of the last four decades in the United States is given in the following table.[23]

Decade	Number of Billion-Dollar Disasters	Associated Cost (billions)	Associated Fatalities
1980–1989	28	$127.78	281
1990–1999	52	$269.68	217
2000–2009	59	$510.38	305
2010–2019	119	$802.08	521

This table shows that the number of events that caused an economic loss of more than 100 billion dollars, the total associated cost in inflation-adjusted dollars, and the number of fatalities is rapidly increasing with each passing decade. There have been numerous devastating events in all regions of the world. Some of them are listed below.

Drought in Southwest U.S.: A severe drought occurred in the Western United States during 2012–2016. This record drought, the worst since record-keeping began, caused a shortage of water in most of California and created a parched landscape in many regions. A study established that anthropogenic GHG emissions increase the probability that higher temperatures will be accompanied by low precipitation—a combination that created this drought.[24] California's Central Valley, which extends for 450 miles from Sierra Nevada to the Pacific Coast, is the single most productive tract of land in the world. A large proportion of fruits, vegetables, nuts, and rice are grown in this region. Since California provides both a substantial amount of agricultural output to the rest of the country and also 15 percent of the agricultural exports from the United States to other nations, a drought in California has a serious impact on the availability of food in the domestic and international markets. This drought caused an estimated loss of $1.5 billion in revenue and of 17,000 jobs. Scientists from NASA, Columbia University, and Cornell University estimated that the chance of a thirty-five-year or longer "megadrought" striking the Southwest and Central Great Plains is over 80 percent if the world stays on its current trajectory of greenhouse gas emissions.[25] Although rainfall in 2017 brought some relief, climatologists fear that low snowfall in the Sierra Nevada mountains may produce another shortage of water in coming years.[26]

High-Intensity Hurricanes: While hurricanes occur in the North Atlantic every year, they have been particularly destructive during the last few years due to strong winds, ocean surges, and enormous rainfalls. Eight of the ten costliest hurricanes on record in the United States have occurred since 2014. As temperatures rise, more water vapor evaporates into the atmosphere, acting as fuel for the storms. Any storm that develops has a greater potential to turn into a hurricane. Modeling predicts a 45 to 81 percent increase in the frequency of Category 4 and 5 hurricanes in the Atlantic coast of North America in coming years.[27] A higher temperature of seawater provides more energy to the storms, thus fueling their intensity. The recorded maximum speed of most powerful hurricanes has also been increasing since 1981.

The 2017 hurricane season was extremely active and deadly. Hurricanes Harvey, Irma, and Maria destroyed thousands of structures and many people were displaced, injured, or killed. Hurricane Harvey dumped 40 inches of rain on Houston, America's fourth-largest city, and displaced 13 million people in Texas, Louisiana, Mississippi, Tennessee, and Kentucky. The damage caused by this hurricane was estimated to be $125 billion by the National Hurricane Center. Irma, a Category 5 hurricane, was one of the most powerful hurricanes in recorded history. The winds blew with a speed of 137 miles per hour for thirty-seven continuous hours and the coastal storm surges were 20 feet above the normal tide levels. It caused the deaths of 102 people, including seventy-five in Florida. The damage caused by Hurricane Maria was estimated to be around $65 billion, a staggering cost for Puerto Rico's ailing economy. Residents had to live without proper food, water, and fuel for many months. Although the official death toll following the hurricane was only sixty-four, researchers from Harvard University estimate that at least 4,645 deaths occurred during the storm and in the following weeks.[28]

Typhoon Haiyan: Typhoon Haiyan, one of the strongest tropical cyclones ever recorded, devastated some regions of Southeast Asia in November 2013. It caused maximum damage in the Philippines, which was hit with sustained winds with speeds up to146 miles per hour. Haiyan caused immense destruction in the country. More than 6,000 people were killed and thousands of homes were destroyed. Over 14 million residents, including nearly 6 million children, were displaced by the typhoon. The financial cost of the damage to the Philippines' economy was estimated to be $14 billion.

Polar Vortexes: In the winter of 2014–2015, states in the Midwest and Northeast had to face brutal winds originating in the North Pole. A polar vortex is a large packet of very cold air, which normally sits over the polar region, driven southward by the flow of winds. A team of American and Korean scientists has established its connection to the loss of Arctic sea ice caused by global warming.[29] They have shown that loss of ice in the Arctic Sea and cold winters in sub-polar regions are dynamically connected through the polar stratosphere. In January 2014, bone-chilling cold, snow, and ice gripped much of the country, affecting about 200 million people. Temperatures in many states dipped to 15° to 25°F. Even in Pensacola, Florida, the temperature was 28°F. Polar vortexes have been a regular feature in recent winters, producing extremely cold weather as far south as Florida.

Heat Waves: Heat waves have also become more common in many parts of the world. A broiling heat wave produced regular temperatures above 110°F in the northern states of India and killed more than 2,000 people in May 2015. The temperature in some parts of India reached 123.8°F, the highest ever recorded. A heat wave that hit Europe in 2003 caused 70,000 deaths. An incredibly intense heat wave hit East Asia in July and August of 2018, with temperatures rising to 106°F in Japan and South Korea. That heat wave is blamed for 119 deaths in Japan and twenty-eight in South Korea. During the same summer, heat waves killed seventy persons in Canada and 180 in Pakistan. Climate scientists believe that such heat waves in many parts of the world are now ten times more likely than in previous decades.[30]

Wildfires: Hot and dry weather caused by global warming has increased the occurrence of wildfires throughout the world. Until a few decades ago, wildfires were primarily seasonal threats that occurred during summer months. Their season has now become longer almost everywhere, and they take place nearly year-round in the United States, Canada, Australia, and other countries. While the origin of increasing wildfires can be traced to global warming, wildfires themselves aggravate climate change because they release copious amounts of carbon dioxide into the atmosphere and destroy forests that constitute an important sink of carbon dioxide.[31] The cost of wildfires in the United States has been estimated to be $665 million per year. In 2017, wildfires torched 199,000 acres of the wine country in Northern California and took 39 lives.[32] Wildfires in the summer of 2018 engulfed more than 200,000 acres in Northern California, destroyed almost 100,000 buildings, and killed eight persons.[33] In August 2018, 109 large forest fires burned forests on more than 1.9 million acres in almost all western states of the U.S. Seven of the most destructive fires on record have occurred in the last eight years. Forest fires in November 2018 were the deadliest and most destructive in the history of the country. A total of 7,579 forest fires devastated an area of 1,667,855 acres (674,957 hectares)—the largest amount of burned acreage recorded in a fire season. More than a thousand homes were burned and close to one hundred people were killed by the fire or while fighting the fires. The summer of 2019 was the worst wildfire season on record for the Arctic, with huge blazes in Greenland, Siberia, and Alaska.[34] A total of 6,872 wildfires were recorded in California that burned 253,321 acres of land. Australia had its worst wildfire season in early 2020 that burned through more than 24.7 million acres (10 million hectares), destroyed homes of thousands of people, and killed more than a billion native animals. The ecosystem may take a very long time to recover from devastation of this magnitude.

Warming of the Arctic: Since 1979, when recording by satellites began, the Arctic has lost more than half its volume of ice, which has diminished in both overall thickness and area covered. Rapid melting of ice has opened shipping lanes across the Arctic, potentially making the Northwest Passage around the North Pole navigable during summer months. Experts have predicted that the Arctic Sea ice will be so thin by the middle of the century that ships will be able to sail directly across the North Pole for the first time. Arctic sea ice plays multiple roles in preserving the ecosystem. It helps cool the entire planet by reflecting sunlight back into space. The loss of sea ice may also adversely affect the life cycle of photosynthesizing organisms at the bottom of the food chain, which support and nourish many species of fish. Additionally, a change in the Gulf Stream will affect the lives of millions of people in the Northern Hemisphere.

Other Incidents: There have been numerous other unusual weather-related incidents during the last few years, increasing in number and intensity each year. An Arctic heat wave at the end of December 2015 caused temperatures at the North Pole to spike to 60° F above normal. Heavy rains in December 2015 caused the Mississippi River and its tributaries to overflow, setting off historic flooding in the U.S. Midwest. A heat wave in India killed at least 2,500 people in 2015. In May 2016, the temperature in parts of Rajasthan, India, soared to an unprecedented 123.8° F. Extremely heavy rainfall in many parts of India in the summer of 2018 caused extensive damage to the infrastructure and led to at least 1,000 deaths. Europe broiled under a record heat wave around the same time. It snowed in Cairo in December 2013, a most unusual occurrence, and there was a 26-inch snowfall in Moscow on April 1, 2013, setting a record. On June 28, 2019, the temperature in a weather station is France was recorded as 114.6°F, which was almost 4°F higher than the previous record.[35] Several western states in the U.S., including Arizona, California, New Mexico, Utah, Nevada, and parts of Oregon and Colorado have suffered from an unprecedented drought that started in 2020 with no end in sight in 2021. It is one of the most severe droughts in more than a few decades and is accompanied by record-breaking high temperatures. It has been suggested that climate warming is accentuating it and making it worse. The drought has already damaged the agriculture industry in California, the state that produces a large proportion of fruits, nuts, and other agricultural products in the country. The drought also increases the risks of wildfires caused by hotter and drier conditions. It has also affected waterways, reducing the generation of electricity by hydroelectric plants.

These are only a few of the numerous extreme-weather incidents that have occurred in recent years. Although any one of these events may be classified as a rare but possible event, the combination of so many events, each of which may occur once in a hundred years, cannot be considered just coincidental and represents a significant deviation from the normal. The large number and high intensity of these extreme events must be attributed to climate change.

Economic Impact of Global Warming

It is impossible to put a price tag on human suffering, loss of lives, and displacement of millions of people due to these events. However, the effect of climate change on the economies of various regions can be calculated. A study found that extreme weather and health effects of the burning of fossil fuels have cost the U.S. economy $240 billion during the past ten years.[36] The average reported losses from weather-related incidents in the U.S. increased from $50 billion a year in the 1980s to $200 billion a year in the following decade. About three-quarters of this total was due to extreme weather caused by global warming.[37] Natural disasters have cost the world's economy $2.5 trillion since 2000. According to projections by the United Nations Development Program, the world's gross domestic product will fall by $33 trillion by 2050 unless drastic steps are taken to decrease the emission of greenhouse gases. Extreme weather-related events have already increased in frequency and severity. This trend is likely to continue, even accelerate, with increasing concentrations of greenhouse gases in the atmosphere.

Sources of Greenhouse Gases

(1) Power Generation

Electricity production is the biggest source of carbon dioxide. Fossil fuel-based power plants use either coal or natural gas as fuel. Coal-fired power plants produce enormous amount of carbon dioxide because the combustion of coal necessarily produces this gas. Natural gas, on the other hand, contains many hydrocarbons; hence it produces energy from the burning of both carbon and hydrogen. While the combustion of carbon produces carbon dioxide, that of hydrogen produces harmless water vapor. In 2019, 38 percent of the electricity in the U.S. was produced from the combustion of natural gas and nearly 23 percent was produced by power plants that burn coal.[38] President Donald Trump, during his tenure, promised to bring back "beautiful, clean coal" power plants with the reasoning that closure of these plants would threaten national security.[39]

The proportion of energy generated by coal-fired power plants is much higher in China (79 percent) and India (68 percent) than in the U.S.[40] It has been estimated that power generation produces 10.8 billion tons of carbon dioxide per year out of a total of 38.2 billion tons of carbon dioxide produced in all sectors. Power plants in China produce 2.84 tons of carbon dioxide, and those in the U.S. produce 2.32 billion tons of this greenhouse gas each year.[41] Besides its easy availability, coal is considered to be cheap because the damage to the health of workers, the general well-being of people living in the vicinity of coal mines and power plants, and its effect on the environment are not taken into consideration.

(2) The Transportation Sector

The second-highest source of greenhouse gases is the transportation sector. In 2018, it accounted for 28.2 percent of the total emissions in the United States. Although carbon dioxide is the main component of the emissions from vehicles that burn gasoline in internal combustion engines, some methane and nitrous oxide are also emitted during their operation. The total emission of greenhouse gases is more from passenger cars and light-duty trucks than from other motor vehicles due to their much greater number. The emission of greenhouse gases from this sector increased by about 16 percent over the 1990–2013 period because of a 35 percent increase in the number of miles driven with these vehicles. The increase in travel miles is attributed to several factors, including economic growth, urban sprawl, and low fuel prices.

The number of vehicles is rapidly increasing in developing countries, with serious implications for the environment. The ownership of vehicles in China is only 128 vehicles per thousand persons, much less than the ownership of 804 vehicles per thousand persons in the United States. However, China surpassed the U.S. and all other countries in the sale of new vehicles in 2010, and the disparity in the ownership of cars is decreasing with each passing year. The number of vehicles per thousand persons was only 18 in India in 2012 but the total number of cars in the country is increasing at the rate of more than 200,000 per month. At present, motor vehicles emit about 670 megatons of carbon dioxide annually in the United States, while emissions from the transportation sector in China and India are 200 megatons and 70 megatons of CO_2, respectively. With the rapid expansion of personal vehicles in these two highly populated countries, their emissions are projected to increase greatly by 2050 if the current trend continues.

Personal cars make a large contribution to the burden of greenhouse gases on the planet.[42] Not only is the transportation sector a large producer of greenhouse gases, it is also the fastest-growing source of CO_2 in the world. Within the transportation sector, urban transportation accounted for almost a quarter of the greenhouse gas emission. As cities continue to grow, particularly in developing economies, these emissions are on track to double in the next 35 years without significant policy intervention. The rapid increase in the number of vehicles in developing countries has alarming implications for the environment. Unless urgent steps are taken to curb the growth of cars in developing countries and to reduce the emission from cars and other motorized vehicles in developed countries, unmanaged growth in motor vehicles alone is projected to exceed the proposed limits on the concentration of greenhouse gases in the atmosphere.

(3) Agriculture

Agriculture contributes to the emission of greenhouse gases in several ways. Industrial agriculture, common in developed countries and becoming the preferred method of farming everywhere, requires a large input of fertilizers and other chemicals on a regular basis. Synthetic fertilizers, insecticides, and weed killers are petrochemical products manufactured from petroleum or natural gas and contain soluble compounds of nitrogen in the form of urea or nitrates.

The portion of fertilizer that is not absorbed by plants eventually breaks down in the soil and produces nitrous oxide gas, which is released in the atmosphere during irrigation and tillage. This gas is highly potent in absorbing solar radiation, making synthetic fertilizer a significant contributor to climate change. The use of fertilizers is rapidly increasing and is the fastest-growing source of nitrous oxide. In addition to fertilizers, insecticides and weed killers are also petrochemical products manufactured from petroleum or natural gas. About 1 percent of total fossil fuel is used to produce 90 million metric tons of these chemicals per year for use in agricultural farms throughout the world. Modern farming methods depend on the use of machines that run on fossil fuels, which also produce carbon dioxide. Rice, the world's second-largest crop, is a source of methane produced by microscopic organisms in rice paddies during the normal growth process of this crop. An increase in ambient temperature decreases the yield, hence increasing the amount of methane emitted per kilogram of rice.

Industrial farming, in general, requires that the farmland be on level ground and clear of all extraneous vegetation. When there is a shortage of arable land, it becomes necessary to raze forests—something that is being done on a massive scale in South America and Southeast Asia. This process adds to the burden of greenhouse gases in several ways. Clearing the land of forests removes an important sink of carbon dioxide because vegetation absorbs a large amount of this gas. Without the cover of vegetation, the fertile topsoil of the forests is washed away—if not immediately then in a few cycles—which necessitates the use of synthetic fertilizers, made of petrochemicals, to grow crops. This method of farming is also detrimental to biodiversity because animals lose their habitat and any vegetation other than the desired crop is considered to be a nuisance and is removed. Industrial farming also requires heavy use of machines that run on petroleum because it is done on large tracts of land and there is a fixed timetable for various operations on the farms. These factors make agriculture an important contributor to climate change.

(4) Livestock Industry

The demand for meat and other animal-based foods is increasing almost everywhere in the world. With increased living standards, people are eating more meat and dairy products. Developed countries already consume a large amount of meat—an average American consumes 214 pounds of meat, 32 pounds of eggs, and 600 pounds of milk or its equivalent in dairy products each year.[43]

With increasing prosperity, the consumption of meat is increasing rapidly in developing countries. Although the per capita meat consumption in China is only about half of that in the U.S., the total meat consumption in China is very large because of its much greater population. The consumption of animal-based foods is increasing at the rate of about 9 percent per year in that country.[44] If the present trend continues, the global requirement of meat will almost double from 2015 levels by 2050.[45]

The increasing demand for animal-based foods has led to the development of Concentrated Animal Feeding Operations (CAFO) in which tens of thousands of farm animals are raised in a single facility. Meat and dairy involve the use of fossil fuels in all stages of production. Their feed is grown on farms with extensive use of fertilizers and pesticides that are petrochemical products. The operation of farms and CAFOs requires the use of diesel oil or petroleum in various machines. Animals are slaughtered in an assembly-line fashion by moving them on conveyer belts and are dismembered with the help of electrical machines. The animals or their body parts are transported hundreds of miles from the beginning of their lives to their end on the dinner table. The requirement to refrigerate and freeze meats and dairy products, in the supermarket or the home of consumers, is more stringent than for plant-based agricultural products. The use of fossil fuels in these operations increases the amount of carbon dioxide in the atmosphere. Since all arable land in the world is already under cultivation, the production of feed for farm animals is often done by razing forests or other vegetation that is not considered useful, thus decreasing carbon dioxide sinks.

In addition to the emission of carbon dioxide from fossil fuels used by the livestock industry, a much greater contributor to global warming is methane gas produced by these facilities. Large amounts of methane are emitted during the fermentation of feed in the first stomachs of cattle and sheep, and also during the bacterial degradation of manure stored in lagoons. Since livestock produce manure at the rate of fifteen to thirty times their weight per year, the amount of manure in these pits is very large, producing a substantial amount of methane. This gas is eighty-five times more potent in causing global warming than carbon dioxide; hence the contribution of livestock to global warming is very large. The total number of cattle, swine, and goats in the world is more than 4.3 billion. When the emission of methane is included, the production of one kilogram of beef produces the equivalent of 34.6 kilograms of carbon dioxide. The corresponding numbers for lamb, pork, and chicken are 17.4, 6.35, and 4.57 kilograms, respectively.[46] During its lifetime, a beef cow produces 4,500 kilograms

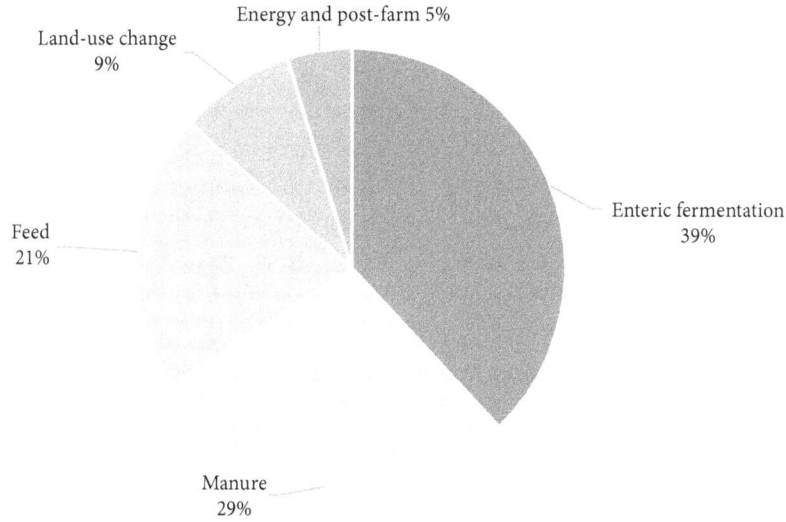

Figure 4 Greenhouse Gases Produced by the Livestock Industry.

of CO_2 equivalent and 40.1 kilograms of sulfur dioxide, which causes acid rain.[47] The distribution of greenhouse gases produced in different phases of the livestock industry, in CO_2 equivalent, is shown in the above pie chart, Figure 4.[48]

The efficiency with which livestock convert agricultural products into food for humans is very low because the animals use a substantial portion of the feed for their growth and sustenance. For this reason, the feed for livestock must contain much more energy and nutrients than the foods produced for human consumption by this industry. Doubling the production of meat and dairy products by 2050, as predicted by extrapolating present trends, can only be done by greatly increasing the agricultural land to grow feed by razing forests. The land area covered by forests has already decreased by almost a third from the pre-industrial era. Razing forests by a substantial amount will have a deleterious effect on the environment because, in addition to providing timber and recreational areas, forests provide a number of essential services, including filtering water, controlling water runoff, protecting soil, regulating climate, storing nutrients, and providing habitat to countless plant and animal species. As the forests are razed to grow feed or to be used as pasture for livestock, the level of greenhouse gases in the atmosphere increases because forests act as sinks for carbon dioxide.

According to the Food and Agriculture Organization (FAO) of the United Nations, the global livestock industry produces greenhouse gases—carbon dioxide, methane, and nitrous oxide—that have a greater warming potential than cars, trains, planes, and ships combined.[49] When emissions from changes in land use are also considered, the livestock sector produces 9 percent of the carbon dioxide, 37 percent of the methane, and 65 percent of the nitrous oxide produced by all human activities. Since methane and nitrous oxide have a much greater global warming potential than carbon dioxide, the contribution of the livestock sector to climate change is very large. Livestock use 39 percent of the Earth's entire surface area, mostly for permanent pastures. In addition, 33 percent of the global arable land is used to grow feed for them.[50] A detailed analysis has shown that forsaking the consumption of animal-based foods will benefit the environment more than giving up the personal car.[51]

(5) Manufacturing

Today's lifestyle requires a very large number of household goods that are produced in factories. The source of power in these factories is either fossil fuels or electricity that is often produced with coal or natural gas. The manufacturing sector makes a large contribution to the concentration of greenhouse gases in the atmosphere. About 22 percent of emissions in the U.S. originate from the manufacturing sector that produces consumer goods for households and businesses. Industrial processes are not limited to factories that produce the final product; all previous steps that convert raw materials into the input of factories also belong to this sector. Some industrial processes require intense heat in a localized area, which is produced by the combustion of fuel in an oxyacetylene flame that produces copious amounts of carbon dioxide. The direct use of natural gas in industrial facilities invariably results in leakage of this potent greenhouse gas into the atmosphere. Manufacturing of plastics, a ubiquitous item these days, also adds to the load of greenhouse gases in the atmosphere.

Consequences of Climate Change

There are many consequences of global warming, some of which are apparent now, while others may have serious consequences with the passage of time. The phenomena that are apparent now include the following:

(1) Decreased Yield of Farms

Perhaps the greatest danger to mankind from global warming is its effect on the production of food. Agriculture has evolved over centuries so that plants remain productive within a narrow range of temperature and humidity. A sudden disruption in this cycle—excessive rain, very little rain, precipitation at the wrong time, or large variations in temperature—can decrease the output of agricultural farms. Climate change is already having a deleterious effect on the production of food on the land and sea, a trend that will accelerate as global warming continues.

Global warming increases the evaporation of water from land, leading to increased drought in dry areas and an expansion of dry areas to neighboring regions. Drought and excessive or inopportune rains have already begun to occur in various parts of the world. For example, the Mississippi River flooded just before the harvest period of many crops in 2009, causing an estimated loss of $8 billion to farmers.[52] During the last decade, severe droughts have affected the agricultural output of many states in the Western United States. Texas had the driest year since 1895 in 2011, and the Southwestern United States suffered from drought and acute shortage of water for many years. California, the most productive agricultural region in the world, had the driest year on record in 2013. In addition to reducing the yield of crops, higher temperatures result in the proliferation of weeds and pests.

The yields of wheat and corn, two of the world's most important staple crops, significantly decrease with an increase in ambient temperature. At a time when the demand for food is increasing due to increasing population and changing food preferences, the production of most grains is projected to significantly decline by the middle of the century. The yield of crops in India, which is the second-largest producer of rice and wheat, is projected to decrease by almost 30 percent by the middle of the century. Agriculture in South Asia is highly dependent on monsoons. Farmers in these countries plant crops according to the timetable of the rainy and dry seasons. A change in this pattern may greatly reduce the output of farms in these highly populated areas of the world. Excessive rain or insufficient rain, as has been occurring in many Asian countries in recent years, decreases the agricultural output of farms. Heavy rain in April in northern India in 2015 and 2016 damaged the winter harvest, causing extensive damage to the crops that were ready to be harvested.

Industrial agriculture that uses hybrid seeds heavily depends on the availability of water at regular intervals. Uncertainty in the availability of water has a greater effect on the output of farms that use hybrid seeds than those that use indigenous varieties. International companies that provide genetically modified seeds, fertilizer, and pesticides encourage the use of hybrid seeds because it makes the farmers financially dependent on them. However, failure of a crop due to bad weather subsequently leaves the farmer heavily in debt. This is one of the reasons for the suicide of hundreds of farmers in developing countries every year. A report from the United Nations climate-science panel states that climate change has already cut into the global food supply and that the trend will accelerate.[53] The world will require 70 percent more food in 2050 due to increasing population and a shift in preference toward animal-based foods. This will cause a dramatic increase in food prices, perhaps resulting in mass starvation. World hunger—a crisis today—will become a catastrophe by 2050 and may result in regional wars and violent conflicts.

(2) Extreme Events

As mentioned earlier, extreme events that have a very low probability of occurrence and should be vanishingly rare are now happening very often. Such events include unusually powerful storms, hurricanes, and tornadoes, extensive periods of drought, heat waves, forest fires, and heavy rains. These events cause extensive damage to infrastructure, hardships to people, economic loss, and even loss of life. Continued emission of greenhouse gases will increase the frequency and severity of such events. According to the German insurance giant Munich Reinsurance America, the overall losses from weather-related catastrophes between 1980 and 2011 were over $1 trillion with an upward trend from year to year.

(3) Health Effects

Climate change endangers human health in many ways, both direct and indirect. The direct effects of climate change on human health are caused by events such as heat waves, droughts, intense hurricanes and storms, and degraded quality of air. Flooding caused by heavy rainfalls takes a toll of life in many parts of the world. According to an estimate by the UN, floods accounted for 47 percent of all weather-related disasters between 2005 and 2014, affecting 2.3 million people and killing 157,000.[54] Heat waves caused a loss of about 150,000 lives during that period, with a large fraction of deaths occurring in high-income countries.

Changes in the environment also adversely affect the health of people in many ways. With warmer weather and changing rainfall patterns, insects that spread infectious diseases increase their range to colder parts of the world. There is already evidence that the mosquitoes that spread dengue fever now live in twenty-eight states of the U.S. The range of Lyme disease, transmitted by the bites of ticks of certain species, may extend to the entire U.S. and Canada. West Nile virus has already spread to many states in the U.S. Warmer climates are also conducive to the evolution of new strains of viruses, such as the Ebola virus. Global warming will increase the occurrence of allergies causing increased human suffering. Allergies and asthma already cost the United States $33 billion annually in direct health costs and lost productivity. With spring arriving ten to fourteen days earlier than it did twenty years ago, pollination of trees is starting sooner. In the fall, ragweed is projected to thrive and become more irritating with increased level of carbon dioxide. Ragweed plants produce twice as much pollen these days than they did a century ago. The pollen production could double again if the level of carbon dioxide in the atmosphere keeps increasing.

Respiratory allergies and diseases may become more prevalent because of increased exposure to pollen, molds, and marine toxins that begin to float in the air. Incidence of cancer is expected to increase due to increased intensity of ultraviolet radiation. Climate change may exacerbate existing cardiovascular disease by increasing heat stress and the number of pathogens to which humans are exposed. Adverse health effects of climate change—both direct and indirect—are increasing every year, affecting the lives of a large number of people in all parts of the world. While economic losses from climate change can be quantified in monetary terms, it is not possible (or fair) to put a monetary figure on the ill-health and loss of life caused by climate change. According to a projection by the World Health Organization (WHO), climate change is expected to cause approximately 250,000 additional deaths per year between 2030 and 2050: 38,000 from heat exposure in elderly people, 48,000 from diarrhea, 60,000 from malaria, and 95,000 from childhood undernutrition.[55]

(4) Rising Sea Levels

The rise in sea levels is often not taken seriously because it is expected to be very small on an annual basis and people tend to ignore dangers that are more than a few years away. However, sea levels are rising continuously and a sudden increase in sea levels is a distinct possibility. Even a small, cumulative increase in sea level can have a devastating effect on coastal habitats. The level of water in the seas had remained essentially constant for many centuries. However, it has been rising at the average rate of about 1.5 to 2.0 millimeters per year in recent decades with large variations in coastal regions.[56] This rate has significantly increased in the last few years. There are two factors that are contributing to rising sea levels: An increase in the temperature of water in oceans increases its volume just as all solids and liquids expand when heated. The other, and more important, reason is the melting of glaciers and polar ice caps. The melting of ice sheets that are currently above water is particularly dangerous because it is a continuous process with the possibility of a sudden increase if a large block of ice falls into the ocean.

The two major ice sheets above water are the Greenland and Antarctic Ice Sheets. These ice sheets are enormous; together they contain 99 percent of the freshwater on the planet. They consist of freshwater and not seawater because they were formed from the condensation of atmospheric moisture eons ago. The Greenland Ice Sheet covers about 656,000 square miles (1.7 million square kilometers). If the entire Greenland Ice Sheet melted, the sea level would rise about 20 feet (6 meters). The Antarctic Ice Sheet is even bigger than the Greenland Ice Sheet since it covers almost 5.4 million square miles (14 million square kilometers), an area that is bigger than that of the contiguous United States and Mexico. Since the Antarctic Ice Sheet is much cooler than the Greenland Ice Sheet, it will melt at a slower rate. If the entire Antarctic Ice Sheet melted, the sea level would rise by about 200 feet (60 meters) with disastrous consequences for all forms of terrestrial life.

Observations have shown that the glaciers in Greenland are melting at a rapid rate. Even a small rise in sea levels may have a devastating effect on coastal habitats, including flooding of wetlands, contamination of aquifers, and lost habitat of birds and plants. As the sea level rises, the damages caused by storms to coastal regions will also increase. There are differences in the rise of sea level in various parts of the world because of the way sea, land, and ice interact. Research published in early 2016 suggests that the eastern United States will experience a greater rise in sea level than the western shores of the country. This rise will gradually upend fisheries in the Gulf of Maine and worsen the risk of damages from storms.[57] In February 2015, scientists at the NOAA determined that the sea level from New York City to Cape Hatteras in North Carolina rose an unprecedented 128 millimeters during 2009–2010.[58] Rising sea levels in the Pacific Ocean have completely submerged five of the Solomon Islands, and six other islands lost at least 23 percent of their landmass. Sea level in the

Solomon Islands is now rising at a significantly greater rate than the global average. Northern Europe, on the other hand, is expected to have a below-average increase in sea levels.

As seawater encroaches the land, it causes destructive erosion, flooding of wetlands, loss of agricultural lands, and displacement of people who live in coastal areas. A sea level rise of three feet, which will occur if half of the Greenland ice sheet melts, will endanger 150 million people and will submerge portions of the world's most populous cities, including New York, London, Tokyo, Manila, Shanghai, and Dhaka. The countries in which more than 5 percent of the population is at risk are the Netherlands, Vietnam, Thailand, Japan, Myanmar, Bangladesh, the United Arab Emirates, the Philippines, Bahrain, and Belgium.

According to the Stern Review commissioned by the British government, about $3 trillion of assets are located less than 3 feet above sea level, including water treatment facilities, power stations, railroads, highways, buildings, and airports.[59] Over 600 million people live in coastal areas that are less than 10 meters above sea level, and two-thirds of the world's cities with populations over 5 million are located in these at-risk areas.[60] Even if cities are not submerged, rising sea levels will increase the damage caused by storms and hurricanes in coastal regions, just as Superstorm Sandy had a crippling effect on life in New York City in 2012. As seawater reaches farther inland, powerful storms can strip away everything in their path and cause extensive damage. Rising sea levels make the storms fiercer and extend the range of destruction to inland areas. According to some estimates, the aggregate flood losses in 136 coastal cities around the world could increase to $1 trillion by 2050.[61]

(5) Acidification and Warming of Oceans

Oceans absorb about one-fourth of the carbon dioxide generated by human activities. The carbon dioxide reacts with water to form carbonic acid, which makes the water more acidic. Increased acidity makes it more difficult for corals, planktons, and other creatures to produce calcium carbonate, the main ingredient in their hard skeletons and shells. A weak shell makes these animals vulnerable to infection and disease. Broader changes in the overall structure of ocean and coastal ecosystem will threaten the survival of marine life—already under existential threat from excessive fishing. Oceans also produce half the planet's supply of oxygen through photosynthesis by algae. Increased acidity will have an adverse effect on aquatic algae, thus reducing the amount of oxygen in the atmosphere.

Warming of the oceans has a side effect that will accelerate global warming. Earth's oceans contain vast quantities of methane in the form of methane hydrate. Colder temperatures keep it stable as an ice-like solid known as permafrost. Warming ocean temperatures, particularly in the Arctic, will melt these substances, thereby releasing methane into the air. It has been estimated that the East Siberian Sea floor will release millions of tons of methane if the temperature of water in the ocean increases significantly. Since methane is highly effective in trapping heat, this release will greatly accelerate global warming.

(6) Melting of Glaciers and Ice Caps

Glaciers and ice caps are natural reservoirs and important resources of fresh water. They increase in size due to additional condensation in winter and release it slowly during summer months. The temperature of water in glaciers is close to the melting point of ice, which makes each glacier an important indicator of global climate in that region. Glaciers maintain the flow of many important rivers during the summer months. The melting of glaciers may greatly decrease the availability of fresh water during the entire year.

Most of the glaciers around the world are shrinking and many of them have disappeared altogether. The rate at which the volume of glaciers is decreasing has been accelerating during the last two or three decades. Glaciers in the Himalayas contain the largest store of water outside the Greenland and Antarctic Ice Caps; these glaciers supply water to seven major rivers that provide crucial support to agriculture and livelihoods in South Asian countries and China. The rivers that depend on glaciers are the Ganges, Brahmaputra, Indus, Yangtze, Mekong, and Yellow. If the glaciers melt, these rivers will become seasonal—they will flow only when it rains on the mountains. This change will deprive hundreds of millions of people of the basic necessity of life. A team of Nepal's scientists found that several glaciers have shrunk by 13 percent in area and the accumulation of glacial ice has become shallower than in earlier times. Several smaller glaciers are now only half the size they were in 1960. It has been estimated that the volume of ice in the Himalayan glaciers will decrease by 70 to 99 percent by the end of the century.[62] This will have drastic consequences on farming, hydroelectric power generation, and the availability of drinking water.

Glaciers are shrinking everywhere, including those in the tropics and at higher latitudes. If all glaciers in Alaska melted, the entire state would be covered by about a foot of water, even though the area of Alaska is 1.5 million square kilometers, greater than the areas of California, Texas, and Montana put together. Alpine glaciers in the Andes are at especially great risk because they tend to be smaller and the tropics are more sensitive to climate change. The melting of glaciers in the high Andes has decreased the flow of the Santa River, creating the possibility of water shortage in Peru. Glaciers in Africa, those atop Mount Kenya and Mount Kilimanjaro, have lost about 80 percent of their surface area.

(7) Loss of Biodiversity

Changes in temperatures and weather patterns will have an impact on both terrestrial and marine biodiversity because many species of plants and animals will lose their native habitats or succumb to changing conditions. The melting of ice in the polar regions is already decreasing the habitat of Arctic and Antarctic creatures such as penguins, polar bears, and puffins. Marine life is in danger because warming of oceans decreases the population of planktons, which are at the bottom of the food chain of marine life. Climate change may cause the extinction of roughly one quarter of all species on land by the year 2050.[63] When the local environment changes, biodiversity provides protection since some subspecies can withstand changes in environment better than others. When conditions are not optimal, these hardy strains will flourish. A rich biodiversity is fundamental to life on the planet.

Recommendations of International Agencies

The United Nations Framework on Climate Change (UNFCC) is the main international organization created to control global warming. The Kyoto Protocol, an international treaty developed by the UNFCC, commits member nations to reduce greenhouse gas emissions. In 2010, parties to the UNFCC agreed that future global warming should be limited to below 2°C (3.6°F) relative to the preindustrial level to avoid the worst impacts of climate change. However, if the world continues to burn fossil fuels at the current rate, global temperature will increase by this amount before 2036. Most scientists believe that even warming by 2°C is already excessive and will have serious repercussions for life on the planet. But we are already hurtling toward a much greater increase in temperature.[64] The report issued by the IPCC in 2015 stated that without policies and actions to mitigate climate change, the expected increase in global mean temperature will be between 3.7° and 4.8°C by the end of the century relative to the preindustrial level. The projections of the American Meteorological Society are similar. Such an increase will trigger widespread disasters in the form of rising sea levels, volatile and violent weather patterns, forest fires, famines, and water shortages. Conflicts over the water of various rivers may result in wars between nations.

The concentration of CO_2 in the atmosphere has been increasing at the rate of 2 percent per year during the last few years, but it increased by a record 3.1 percent in 2016 and by more than 2.5 percent in 2017.[65] According to a very conservative estimate, the concentration of CO_2 in the atmosphere should not become greater than 450 ppm if the object is to limit global temperature rise to 2°C.[66] To avoid this threshold, global carbon emissions could rise only for a few more years and then would have to ramp down by several percent a year. Achieving this goal will require a 40 to 70 percent reduction in emissions by 2050 compared to 2010 levels, and zero emissions by the end of the century. The accounting firm PricewaterhouseCoopers analyzed the present trend, the goals stated by various countries, and the reductions in carbon required to keep the global temperature increase below 2°C. According to the firm's calculations, the world will experience at least a 4°C temperature increase. Keeping global warming below 2°C would necessitate that all economies double their goals of carbon reduction in a very short period.[67] The Stern Review also concluded that, in the absence of rapid and concrete actions, there is a 50 percent chance of a rise by 5°C, which is "very dangerous indeed."

Despite these dire predictions from climatologists and other scientists, the governments of major countries have been slow in taking serious steps to deal with the climate crisis. The UNFCC established a framework for action for the 193 members of the UN aimed at stabilizing atmospheric concentration of greenhouse gases. Since 1995, world governments have met every year to forge a response to climate change. Although most nations agree in principle to reduce their greenhouse gas emission, the progress in the last 25 years has been minimal and many countries have made only vague promises.[68] China promised to cut GHG emissions but the Global Energy Monitor found that the country increased its coal-burning capacity by more than 40 gigawatts in the 18-month period up to June 2019. President Trump declared in June 2017 that the U.S. was pulling out of the Paris Accord, stating that it would cost $3 trillion to the United States and result in the loss of 6.5 million jobs. The withdrawal was finalized in November 2020. After Syria signed the Paris agreement, the U.S. became the only country in the world that has excluded itself from the agreement. The U.S. will not provide the $2.7 billion promised by Barack Obama's administration to help developing countries reduce their dependence on fossil fuels. After Trump declared that the U.S. was pulling out of the Paris Accord, eleven states in the country, including Washington D.C. and Puerto Rico, vowed to pursue policies that will uphold the commitment of the United States to the Paris Accord. They have joined to form a bipartisan group called the United States Climate Alliance, which seeks to reduce greenhouse gas emissions nationwide. In addition, 30 cities and 80 university presidents have agreed to work toward the goal of reducing the emission of greenhouse gases in compliance with the Paris Accord. Despite the withdrawal of the United States from the Paris Accord, it is likely that most countries will reduce their emission of greenhouse gases, although not to the level required by climatologists. On the very first day of his administration, President Joe Biden announced that the United States would rejoin the Paris Climate Accord.

Unless governments, industries, and individuals take drastic steps to curb the emission of greenhouse gases soon, we are heading toward events that will be highly destructive, even catastrophic, for the human race. The documented events caused by climate change, along with many others yet to be discovered and analyzed, threaten the welfare and the continued survival of large sections of humanity. The irony and cruelty of climate change is that it will have a devastating effect on the poorest countries that have limited resources and have not significantly contributed to global warming. Those with ample resources in the developed world or in developing countries can insulate themselves to some extent from the vagaries of weather and uncertainties in the availability of provisions.

Mitigation of Climate Change

According to the IPCC and other authorities, the atmospheric concentration of CO_2 must be stabilized at 450 ppm to have a fair chance of keeping global warming below 2°C. Even though other gases also contribute to global warming, the expectation is that methods to control CO_2 will also reduce other greenhouse gases such as methane and nitrous oxide. Stabilizing CO_2 at a certain level requires not only stabilization of its emission, but also its reduction to a level at which natural processes can reduce its concentration.

Energy Supply

Since coal-fired power plants are the biggest source of CO_2, the most important step is to replace these plants with those that produce no CO_2 or at least
minimize its amount in the atmosphere. The plants that cannot be shut down
soon due to lack of alternatives should be fitted with carbon capture and storage technologies that transport this gas to a storage site and deposit it in an
underground geological formation, so that it is not released into the atmosphere. Power plants that use natural gas as fuel also emit a substantial amount
of greenhouse gases and must be phased out. In addition to an emphasis on
solar energy, it is also important to develop and invest in alternate technologies
such as wind, geothermal, and tidal waves.

Transportation

Personal cars make a large contribution to the global emission of greenhouse
gases. Urban transport causes 670 megatons of CO_2 emission in the United
States per year.[69] Unless Americans moderate their love affair with cars, and
the developing economies do not emulate the U.S. model, "cars will cook the
planet."[70] In order to mitigate the GHG emissions from cars, governments
must invest heavily in public transportation, both to make it easily accessible
and to reduce its cost. Investing in public transportation and making it easier
for people to walk or use bicycles will save a large amount of public and private
capital. Studies have shown that public transportation is beneficial to individuals, communities, and the world. Investing in public transportation and educating people to use bicycles in urban areas will more than halve the emission
of greenhouse gases from the major cities of the world.

The overwhelming amount of CO_2 produced by cars can be explained by
a simple calculation. Each gallon of gas, weighing about six pounds, produces
about 20 pounds of carbon dioxide because the ignition of gasoline picks up
large amounts of oxygen from the air. This amount of carbon dioxide, under
normal atmospheric conditions, will fill a balloon with a diameter of 41 inches.
With millions of cars on the road at any time, urban areas are figuratively
submerged in this ocean of balloons. Any reduction in the consumption of
gasoline, either by driving fuel-efficient cars or decreasing the mileage driven,
reduces the burden of GHG in the atmosphere. Estimates have shown that
the development of clean public transport and non-motorized vehicles could

reduce emissions by roughly 1.7 gigatons of CO_2 per year.[71] In the emerging economies of China and India, people use personal cars for transportation even when bicycles would be more useful or where public transportation is available because traveling by car is a status symbol. This makes educating the public very important to reduce climate change.

Industries and Municipalities

Industry uses about a quarter of the energy produced in the world, resulting in the emission of 13 gigatons of CO_2. Businesses of all sizes can reduce emissions by increasing the efficiency of motors and eliminating air and steam leaks. Energy efficiency should be a major consideration in the design of new facilities. Motors running on gasoline or diesel fuels should be replaced by electrical motors. Municipal solid waste and wastewater contributed 1.5 gigatons of CO_2 in 2010. Improvements in energy and waste treatment technologies could result in significant reduction of GHG from this sector. Green buildings should be built with processes that are environmentally responsible and resource-efficient in all phases of construction and maintenance.

Agriculture, Forestry, and Other Land Uses

Agriculture and the conversion of forests to farmlands or pastures cause the emission of about 10 gigatons of CO_2 equivalent per year. Reducing deforestation and planting new forests is the most cost-effective way of reducing the emission of GHG from this sector. The industrial method of raising livestock increases the burden of greenhouse gases in the atmosphere in various phases of operations. It has been estimated that the global warming directly attributable to livestock was 23 percent of the total in 2010.[72] Drastically reducing the consumption of animal-based foods will significantly decrease the concentration of greenhouse gases in the atmosphere, thereby slowing climate change.

Carbon Sequestration by Organic Farming

Organic farming is significantly different from industrial farming. Since it does not use synthetic fertilizers, the soil contains substantial amount of humus—the organic matter that is formed in the soil from the decay of leaves, twigs, stems, etc. Such soil is more porous and has a greater capacity to hold water.

In addition to organic matter, good soils are inhabited by diverse populations of earthworms, insects, arthropods, and microorganisms.[73] Interaction
between the inert soil, organic matter, and these minute creatures constitutes a
dynamic system that preserves and maintains the soil's fertility. While nitrogen
in a bio-available form is provided by synthetic fertilizers in conventional farms,
it is integrated in the soil in organic farms by growing legumes, recycling crop
residues, and composting the manure of livestock. Soil-building practices such
as crop rotation, intercropping, cover crops, and minimum tillage are central
to organic practices. Organic agriculture takes a proactive approach to treat
problems before they emerge.

Organic farming produces lesser quantities of greenhouse gases because it
does not use fertilizers and pesticides produced from fossil fuels. In addition to
contributing to global warming by the emission of CO_2, the chemicals used in
conventional farms also contaminate the groundwater. Since no synthetic fertilizers are used, the highly potent greenhouse gas, nitrous oxide, is not emitted
from organic farms. As an added benefit, a meta-analysis of published literature
has shown that organically grown fruits, vegetables, and crops have more antioxidants and lower concentrations of undesirable heavy metals as compared to
conventionally grown produce.[74]

Organic farming sequesters carbon dioxide in the soil, thus reducing the
amount of this gas in the atmosphere. The organic matter present in the humus
represents carbon that was captured from the atmosphere by plants when they
were growing. Once it becomes part of the topsoil, the carbon will stay there,
thus decreasing the load of carbon dioxide in the atmosphere. Several studies
have shown that due to mineralization and sequestration, the global warming
potential of organic farms is lesser than that of conventional farms by 26 to
80 percent.[75] Studies have also shown that the soil in organic farms have two
to three times the organic matter than that of conventional farms. A greater
concentration of organic matter boosts microbial activities which benefits soil
conservation and plant growth. With proper management, organic farming
may even become a net absorber of CO_2.

Economic Considerations

The steps necessary to mitigate climate change—building power plants that do
not use fossil fuels, developing public transportation facilities, making green
buildings, expanding land for afforestation—require an investment of capital.

However, if nothing is done to minimize and reverse the accumulation of greenhouse gases in the atmosphere, humanity has to live with decreased yields of farmland, droughts, floods, wildfires, hurricanes, and other extreme events that will continue to increase in intensity, causing greater and greater damage with each passing year. Climate change also adversely affects the health and life cycles of people, which cannot be quantified in economic terms.

Calculation of the cost of not doing anything to reverse the accumulation of greenhouse gases involves extrapolation of the present trend for the next few decades. This is done with the help of data from the last few years using the best scientific knowledge to forecast the future. Although there are some differences in estimates of the costs involved, all analyses indicate that if no action is taken soon the cost of mitigating climate change will increase exponentially. At a certain stage, the damages caused by climate change will be so large that reversing the trend will be very difficult and expensive.

The IPCC produced a report representing the work of 1,250 experts from all over the world that concluded catastrophic climate change can be averted without sacrificing living standards. Diverting hundreds of billions of dollars from fossil fuels into renewable energy and eliminating wastage of energy would decrease the expected annual growth rate by only 0.06 percent.[76] The report's conclusion was that tackling climate change is economically affordable, that carbon emissions will ultimately have to fall to zero, and that these steps will help to alleviate global poverty.[77] The cost of limiting and reducing the concentration of greenhouse gases will be much greater if needed actions are not taken soon and postponed for a future date. The IPCC states that mitigating climate change will trim economic growth by a tiny fraction and may improve growth by providing jobs in new industries and reducing healthcare costs. The Stern Review also makes the case for urgent global action because climate change presents serious global risks everywhere. If no action is taken to decrease greenhouse gas emissions, climate change will continue to reduce the global GDP by at least 5 percent each year. In addition to the direct cost of grappling with extreme events, climate change could have serious impacts on the growth and development of the economy. In contrast, the cost of action—reducing greenhouse gas emissions to avoid the worst impacts of climate change—can be limited to around 1 percent of global GDP each year.[78]

A study by the International Monetary Fund (IMF) concluded that reducing the emission of greenhouse gases through carbon pricing (a tax on the carbon content of fossil fuels) will raise revenues that would permit tax reduction in other areas and will be beneficial for most countries. The combustion of fossil fuels, especially coal, is a leading cause of air pollution which, according to the WHO, is estimated to cause over 3 million premature deaths a year in the world, and an increase in the risk of heart disease, lung cancer, etc. Taxing the carbon content of coal will increase its price and decrease its use, leading to lower CO_2 emissions, better public health due to cleaner air, and reduced expenditures on health services.[79] In short, mitigating climate change is eminently affordable and will also provide several ancillary benefits.

Transition from an economy that makes extensive use of fossil fuels to a zero-GHG-emission economy will change the labor market because some jobs will be eliminated while others will be created. The major sources of greenhouse gas emissions—power plants that produce electricity, private transportation, equipment for heating and cooling residential and commercial buildings, agriculture, and industries—will have to make considerable changes to reduce, or even eliminate, their dependence on fossil fuels. The industries that manufacture petrochemical products, such as plastics and fertilizers, will have a reduced workforce, but there will be a need for workers in new industries—for example, those that retrofit pollution control devices in existing outfits, manufacture rails and associated equipment, and organic farming. A sustainable economy will, on the whole, create more jobs, although their required skills will be different. A reduction in the emission of greenhouse gases will save lives and decrease the incidence of diseases caused by the combustion of coal. A reduction in extreme climatic events will reduce human suffering in all parts of the world.

Opposition to Climate Change

A survey by Pew Research in 2017 found that about 61 percent of people in the world believe that climate change is a very serious problem for which some action must be taken soon. The concern about climate change in the United States, which has very high per capita emissions, is somewhat lower at 56 percent.[80] In this country, there is a stark difference of opinion between Democrats and Republicans. While 68 percent of Democrats believe that climate change is a very serious problem, only 20 percent of Republicans hold this view. People living in Africa, Latin America, and Asia, many of which have very low emissions per capita, are the most concerned about the negative effects of climate change.[81] The reasons for the opposition to climate change mitigation by the GOP includes its aversion to any changes in the status quo, and the fact that GOP leadership receives large amounts of campaign contributions from oil, gas, and coal industries.[82] The leadership of GOP also keeps the interests of businesses ahead of anything else.

Industries are in general opposed to actions that may be taken to mitigate climate change for several reasons. Companies that control fossil fuels, such as Gazprom, Exxon Mobil, Shell, and Shanxi Coking, consider a shift to a sustainable economy a threat to their very existence. The same is true of industries that manufacture petrochemical products. Facilities that depend on fossil fuels for their operations oppose mitigating actions because the necessary changes will require major expenses. Many establishments will spend substantial capital only if they are forced to do so by regulations or when the government provides the finances to make the change. Many businesses are also threatened by moves toward sustainable living that involve behavioral and lifestyle changes because these changes may disrupt their operations and profits.

Industries that view a switch to a sustainable economy as an existential threat try their best to mold public opinion. These interest groups search for any discrepancies in the scientific arguments, even those that have been disproved by later scientific work, and promote them relentlessly in the media to convince the public that the scientific case for global warming is not firmly established. A study by Drexel University's environmental sociologist Robert Brulle analyzed the source of funding of groups that promote climate change denial.[83] From 2003 to 2007, the ExxonMobil Foundation and Koch Affiliated Foundations provided financial support to many organizations that deny climate change. ExxonMobil is the world's most profitable corporation that makes most of the money from oil. Koch companies and their affiliates are involved in refining and distribution of oil. These companies have a lot to lose from efforts to tackle climate change. To safeguard their profit, they need to discredit the scientific evidence in favor of climate change—which they do by supporting those scientists who speak in scientific jargon and oppose the notion of climate change caused by human actions.

A total of 140 foundations funneled $558 million to almost one hundred climate change denial organizations from 2003 to 2010.[84] Those scientists who deny the role of human-induced greenhouse gases in global warming pick up selections from scientific research and, with the support of funding from organizations clinging to the status quo, repeat them in the popular media. One such person is Willie Soon, a researcher at the Harvard-Smithsonian Center for Astrophysics, who received $1.25 million from ExxonMobil, the Charles G. Koch Foundation, and the American Petroleum Institute.[85] Soon, with his Harvard-Smithsonian credentials, was frequently held up as an authority by those who reject the underlying science behind climate change and was sought after and admired by climate change deniers. Soon did not receive any grants from NASA or the National Science Foundation, the agencies that usually fund research on climate science. In 2015, the Smithsonian and several academic journals started an ethics investigation against him for his failure to report that his work was funded by energy companies.[86]

Since 2008, most funding for denial efforts has come from untraceable sources. Fossil fuel companies have distanced themselves from open climate change denial and have begun directing their support through secretive networks such as the Donors Trust. Peabody Energy is one of the largest coal-mining companies in the United States. Its funding to groups that deny climate change became known only when it was forced to seek bankruptcy protection due to competition from cheaper natural gas. It provided financial support to numerous groups opposing climate change action that do not want any environmental regulations on polluting industries. These groups include trade unions, conservative think tanks, and corporate lobby groups.[87] Peabody also supported contrarian scientists such as Richard Lindzen and Willie Soon.

Some academics are attracted by the prospect of getting substantial funds and, with their scientific backgrounds, create arguments that sound logical to the general public. Political and financial support from wealthy organizations amplifies the voices of the dissenters on climate change even when they do not represent the scientific community. Politically conservative newspapers and tabloids like the *Wall Street Journal, Daily Mail, Telegraph*, and *The Australian* spend a disproportionate amount of time and space in amplifying the voices of less than 3 percent of climate-science contrarians, as well as nonscientists with similar views.[88]

Another weapon in the arsenal of those who oppose any action to mitigate climate change is to hand-select a few words from climate scientists' statements, rephrase those words to mean something entirely different, and then criticize the scientists for saying it.[89] The neutral media, in order to show that their approach is even-handed, give equal time to both legitimate climate scientists and those who have been propped up by the energy companies. This approach helps the contrarians by giving them a platform to publicize their viewpoints. Since the source of funding of all participants is neither disclosed nor investigated, a nonscientific observer is left confused and may simply choose the path of least resistance: to do nothing for the time being.

President Trump fiercely opposed the notion of climate change. Many of his Cabinet members also deny climate change. Scott Pruitt, when he was the EPA chief, removed all mention of climate change from the government websites and replaced scientists from the advisory panel with supporters of the fossil fuel industry. In July 2018, Scott Pruitt was replaced by Andrew Wheeler, who pursued the same policies. President Trump imposed a 30 percent tariff on solar panels to make them less competitive with the fossil fuel industry. His administration ended NASA's Carbon Monitoring System, a $10 million per year effort to monitor global carbon emissions. This move jeopardizes plans to verify the national emission cuts agreed to in the Paris climate accord.[90] In a stark reversal of Trump's policies, President Biden has elevated the issue of climate change across the U.S. government and tried to accelerate the nation's shift away from fossil fuels. His administration has paused oil and gas leasing on public lands and waters, and has pledged to cut the emission of greenhouse gases by 50 percent by the end of the decade, a commitment that will require major changes in the ways Americans live, work, and travel. President Biden announced that the U.S. is rejoining the Paris climate agreement and revoking the Keystone XL oil pipeline's federal permit. He also pledged to review the regulatory actions taken by Trump to support high-emitting industries.

The American Meteorological Society predicts that unless steps are taken soon to reverse the accumulation of greenhouse gases, there is a 90 percent probability that global temperatures will rise 3.5° to 7.4°C by the end of the century. The consequences will be devastating in many parts of the world and will endanger the lives of people everywhere. Humans have been using fossil fuels with abandon to obtain numerous comforts and conveniences while the adverse effects of the accumulation of greenhouse gases keeps multiplying. According to James Hansen, a retired NASA climate scientist who led research showing that dangerous climate shifts will take place within decades not centuries: "We are in danger of handling young people a situation that is out of their control." The awareness that their future may be compromised is slowly percolating to the young people. Most American teenagers and many young people around the world believe that the climate change will cause harm to them personally and to other young people.[91]

The Way Forward

Climatologists, scientists, and some global leaders have been making a clarion call to take immediate action to reduce greenhouse gases for the last few years. With each passing year, the cost of preventing the consequences of further climate change grows exponentially. The task of slowing—and reversing—the emission of greenhouses gases is so huge that it can only be achieved through concerted actions by individuals, businesses, and governments. Since a considerable amount of time has been lost in discussions and foot-dragging, it is imperative that actions be taken expeditiously. Businesses have been particularly averse to taking steps to mitigate climate change because any movement that comes in conflict with continuous growth and increasing profits will decrease their financial base. In many countries, governments pay more attention to moneyed interests than to the general public and are reluctant to take steps that may alienate this powerful group.

Individuals have a major role to play in slowing down and reversing the accumulation of greenhouse gases. To some extent governments and businesses are beholden to citizens and consumers. An educated and committed citizenry can remove the roadblocks and force actions by both these bodies. The very first step is to believe in the scientific evidence and not be distracted by pseudo-science masquerading as genuine science. Climate change is real, it is happening now, and is getting worse with the passage of time. The only way to stop its progression, and to possibly reverse it, is by decreasing the use of fossil fuels—both direct and indirect. This will be an uphill battle against many industrial houses. They will keep saying, with the megaphone provided by media, that there is no such thing as climate change and, even if there was, human activities have nothing to do with it and hence things should continue as usual.

Businesses, in general, will be opposed to any actions on climate change because any action to reduce greenhouse gases will change the status quo and adversely affect their profits. Most businesses work with a short time horizon because they must achieve growth on a regular basis. They will take actions to reduce global warming only when it helps the bottom line or improves their image with the public. For this reason, some major corporations are endorsing and accepting changes that will benefit the environment. Walmart, Chevron, GE, and many other corporations now want to embrace the green label. However, since corporations always act in their own interests and seek to maximize profits, it would behoove consumers to make sure that their interest in preserving the environment is not a public relations ploy without substantive actions.

In democratic societies, governments must come periodically to people with an agenda to ask for a mandate. It is up to the citizens to make sure that the environment and steps to be taken to combat climate change remain on the agenda, and that each candidate must state his or her position on this issue, including steps that will be taken to reverse the present trend. The problem is that, at the time of elections, several subsidiary issues crop up—or are made to crop up—that divide the electorate. Voters should make a list, along with priorities assigned to each item, to decide the candidates to be supported and hold them responsible for their positions when they are in power.

Fossil fuel companies, just like most other companies, are publicly traded. Many investment houses, educational institutions, or other groups have financial stakes in them through equities. Members or investors can demand that these institutions divest from industries that support fossil fuels. Institutional investors control a substantial portion of the market. Withdrawal of their support will put pressure on the industries that profit from the use of fossil fuels. All of this makes it very important to have an informed citizenry that is willing to take actions within its powers and keeps reminding the policymakers of the importance of the issue. The movement to divest from fossil fuel companies is gaining momentum. In December 2017, Governor Cuomo announced that New York will divest its vast pension fund investment from fossil fuels, French President Emmanuel Macron has decided not to give any licenses for oil and gas exploration, and the World Bank has said it will no longer lend money for the exploration of fossil fuels.[92]

Our personal carbon footprint is determined by the things we buy, our mode of transportation, and our diets. The demand by consumers for goods and services that do not contribute to the accumulation of greenhouse gases will be a key driver in the movement to arrest climate change. This awareness has already begun to modify consumer preferences. The products that consumers buy everyday have a carbon footprint that can vary by substantial amounts. Buying articles that originate in distant lands will add to the burden of greenhouse gases in transportation, in addition to the emission of these gases in the production process. A change in means of personal transportation is only possible when alternate public transportation is easily available. Pressure from the public is necessary to prod authorities to provide mass transit. However, buying fuel-efficient cars and reducing unnecessary driving will also be helpful.

Another important development that will make a substantial difference in the atmospheric accumulation of greenhouse gases is organic farming. The essential requirement of organic farming is a soil rich in organic matter, which provides nutrients to the plants without the use of synthetic fertilizers. Since fertilizers, herbicides, and pesticides are not used in organic farms, the use of fossil fuels is much less than in conventional farms. As a bonus, the soil of organic farms stores carbon in the form of organic matter, thereby reducing the amount of greenhouse gases in the atmosphere. From an operational point of view, organic farming must be small-scale because farmers cannot follow a standard recipe for the application of water and chemicals at times predetermined by the seed companies but must take corrective actions as required by the situation. For this reason, organic farmers must be more knowledgeable than those involved in conventional farming. However, the profits from the sale of organic produce are almost twice as much as those from conventional produce. The sales of organic foods are growing at the rate of 9 percent per year with projections that this rate will further increase with the popularity of organic foods. These factors will encourage entrepreneurs to enter this field because it is an industry that has a potential for substantial growth.[93]

An argument often advanced by those who oppose any action to prevent climate change is that developed countries should not reduce the emission of greenhouse gases until developing nations reduce their emission. However, it is useful to remember that residents of developed countries consume much more energy than those of developing countries. For example, the average per capita energy consumption in the U.S. is five times that of an average person in China, and sixteen times that of an average Indian.[94] There are millions in those countries who do not have the basic amenities. It is possible for Americans to reduce their GHG usage by substantial amounts because their consumption is so large. The author Naomi Klein writes: "Climate change deniers like to claim that environmentalists want to return us to the Stone Age. The truth is that if we want to live within the ecological limits, we should need to return to a lifestyle similar to the one we had in the 1970s, before consumption levels went crazy in the 1980s."[95] Earth's atmosphere is a global resource. Humans have proved that they are very intelligent by making fantastic instruments that make our lives comfortable and give us enormous power. Now we must show that we are also capable of collective action to save the future of humanity. Mitigating climate change is possible but we need actions to implement ideas soon.

TWO

· · · · · · · · · · · · · · · · · · · ·

WATER: THE MOST PRECIOUS RESOURCE

Water is a precious resource essential for the survival of all forms of terrestrial life. We need water for direct consumption and personal hygiene on a regular basis. Of all human activities, agriculture requires the greatest amount of water. The number of people the ecosystem can support is determined by the available amount of fresh water. The Earth contains a large amount of water but more than 97 percent of it is in the oceans. The high concentration of salt in seawater makes it unsuitable for direct consumption by all forms of terrestrial life. A large portion of fresh water is locked in glaciers, in permanent ice caps at high altitudes, and in regions around the North and South Poles. Major sources of water on which civilizations have depended for millennia are rivers, lakes, and swamps. Groundwater, another source of fresh water, is the water that has been stored in the soil or in rock formations. An accumulation of a substantial amount of groundwater that can be extracted using a well is known as an aquifer.

The ultimate source of fresh water on the Earth is energy from the sun. Solar energy causes water to evaporate from oceans and forms clouds that contain water in a pure and salt-free form. Clouds are transported by winds and provide water to the terrestrial regions through rainfall, recharging the aquifers and increasing the volume of frozen snow in glaciers and permanent ice caps. Snow and ice in glaciers and mountain peaks play an important role in maintaining the flow of water in rivers throughout the year because the molten snow provides water in summer months. Without these sources of water in the frozen form, most rivers would become seasonal with large variations in the flow of water during rainy and dry seasons.

The demand for water is increasing for many reasons, including urbanization, higher standards of living, increasing population, and changing food preferences, which all greatly increase the demand for water. A dietary shift to meat and dairy increases the demand for water because the production of animal-based foods requires much more water than the direct consumption of agricultural products by humans.

Droughts and irregular rainfalls caused by climate change also interfere with the availability of water. These developments have stretched the need for fresh water to the point that shortages are projected in many countries within the next few decades. There are already serious, dangerous shortages in many parts of the world. Around 1.2 billion people currently face water shortage daily. Another 1.6 billion people suffer from the consequences of an insufficient supply of pure water, such as malnutrition and disease. Water shortage in several Southwestern states in the U.S. is a serious problem because it decreases the productivity of farmlands. A government report found that there may be water shortages in 40 states of the country within a decade or two.[96] There are two main reasons for the projected shortage of water in the country. The first is population growth, which increases the demand for water and food. Second, the amount of water is expected to decrease due to evaporation from reservoirs and streams as the climate gets warmer. A problem somewhat related to the shortage of water is that of pollution. In areas with substantial industrial, agricultural, or livestock activities, available water often becomes polluted and unsuitable for consumption. According to the U.S. EPA, 35 percent of rivers and 45 percent of lakes, reservoirs, and ponds are polluted in the country.[97]

Pollution in these bodies of water makes them unsuitable for direct consumption, swimming, and the growth of aquatic life. About 80 percent of wastewater is dumped—largely untreated—back into the environment, polluting rivers, lakes, and seas. Water gets easily polluted because it is a good solvent and dissolves more substances than any other liquid. This is also the reason that it gets easily polluted. Toxic substances and waste materials from farms, towns, and factories readily dissolve in water, causing water pollution.

Global Water Shortage

There is already an acute water shortage in many parts of the world. Freshwater consumption has more than doubled in the last few decades and is increasing at a steady rate. Water is becoming scarce globally, and every indication is that it will become even more scarce in the coming years. The problem of a lack of water for daily necessities is growing at an alarming rate and is beginning to affect almost all countries of the world. More than a billion people in the developing world have no access to reliably clean drinking water. Climate change is causing droughts and reducing the volume of water flowing in rivers, while the demand for water is increasing rapidly. Urbanization and increasing demand for energy also put stress on water resources because traditional electrical power plants require copious amount of water for their operation. According to a UN report released in March 2015, unless people everywhere drastically reduce their water use, the world could suffer a 40 percent shortfall in its availability in just fifteen years.[98] The world faces a water crisis that will touch every part of the planet; a point that has been stressed by Jean Chrétien, former Canadian prime minister and co-chair of the InterAction Council: "The future political impact of water security may be devastating. Using water the way we have in the past will not sustain humanity for a long period."[99]

Shortage of water immediately translates into shortage of food because agriculture requires large amounts of water—the productivity of farmlands is determined by its availability. The output of farmlands in countries that produce large amounts of food, such as India, China, Australia, and Spain, is already limited by the amount of water available. The flow of water in some of the world's most important rivers, including the Niger in West Africa, the Ganges in South Asia, the Yellow in China, the Danube in Europe, and the Nile in Africa, has declined precipitously due to overexploitation, mostly for agricultural purposes. The Nile and Yellow rivers no longer reach the ocean most of the year because a large amount of water is drawn upstream.

The band of countries stretching from China, India, Pakistan, and the Middle East to North Africa is severely water stressed and may soon not have enough water to maintain the current level of food production.[100] The World Bank has predicted that the economies of the Middle East, Central Asia, and Africa will be severely affected by the middle of the century due to shortages of water, which may even lead to mass starvations.[101] Even these days, around 4 billion people—almost two-thirds of the world's population—face water shortages for some period of each year. In this age of global interconnectivity, water scarcity and the consequential food scarcity will have an impact on the entire planet because people everywhere consume food imported from regions that may be affected by climate change.[102]

There is already a shortage of water in some parts of the U.S. and, according to the EPA, 40 of the 50 states are expected to have water shortages in some regions in just a decade or so.[103] According to this report, one state (Montana) has a state-wide water shortage right now and 24 states, mostly in the West or Midwest, have regional water shortages.[104] These states are WA, OR, CA, NV, ID, WY, SD, NE, CO, NM, TX, OK, MN, WI, Il, MI, AR, LA, MS, FL, and NC. In addition, the following 15 states may have local water shortages in the coming decades: AK, AZ, KS, MO, TN, SC, VA, WV, PA, NY, RI, MA, NH, ME, and NJ.

In Florida, there is an impending shortage of water, primarily due to increasing population. The level of water in the reservoirs in upstate New York has been dropping almost every year. The amount of water in Lake Mead, the reservoir that stores water for parts of Arizona, Southern California, Southern Nevada, and New Mexico, decreased to a record low in July 2021. The Colorado River is one of the most important water sources in the United States. From its source high in the Rocky Mountains, the Colorado transports water nearly 1,500 miles to the delta in Mexico. It provides water for municipal, industrial, and agricultural use in Wyoming, Colorado, New Mexico, Utah, Nevada, Arizona, and California. It supports the lives and livelihoods of about 40 million people and helps irrigate 3.5 million acres of cropland that produce 15 percent of the nation's food supply. Its other uses include hydroelectric power generation and support for fish and wildlife. The flow of water in the river has been decreasing for the last few decades. Brad Udall, Senior Water and Climate Research Scientist at the Colorado Water Institute, estimates that the flow of water in the river will substantially decrease during the coming years due to lower precipitation in the Rocky Mountains.

Fresh Water and Climate Change

Higher temperatures caused by climate change increase the loss of water by evaporation.[105] The present and projected shortages of fresh water will become worse with higher temperatures and other deleterious effects of climate change. As temperatures increase, crops, farm animals, and all living things require more water, hence climate change is likely to increase the demand for water while shrinking the supply. Lesser amounts of snow in glaciers and snow-covered peaks will reduce the flow of water in rivers that originate from those regions during summer months. Changes in precipitation patterns will create drought in some places, and excessive rainfall in others. Heavy rainfall exceeds the capacity of land to absorb water; the excess then flows into rivers and seas. Due to these changes, the existing infrastructure will not be able to deliver adequate water to the population throughout the year in many parts of the world.

Water Usage

A typical Westerner uses about 80 to 100 gallons of water every day for cooking, showering, flushing toilets, and other household uses. The amount of water used for daily activities greatly increases with greater affluence because people require running water twenty-four hours a day, better sanitation services, and also water for many other activities such as watering lawns and gardens, washing cars, and indulgences such as Jacuzzis and private swimming pools. The trend toward urbanization occurring everywhere in the world requires an infrastructure to deliver water to individuals and to process the wastewater from homes and businesses.

A family of four in a developed country uses about 400 gallons of water every day. A typical distribution of domestic consumption is shown in Figure 5. The arid West has some of the highest per capita residential water use, primarily for irrigation of home lawns and landscapes. An average American uses almost seven times the water used by a Chinese person. The discrepancy is even greater

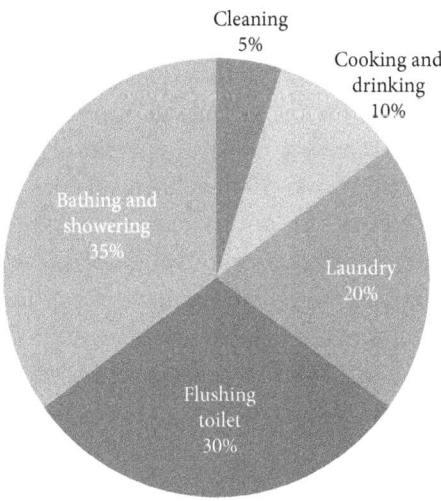

Figure 5 Domestic Consumption of Water in the United States.
Source: EPA.

if one compares the water consumption in developed countries with the consumption in the countries of Africa and Middle East.[106] Water is also required for manufacturing consumer products such as paper, clothes, appliances, and other goods. However, domestic water consumption is greater than industrial water consumption in the United States.

Of all human activities, agriculture requires the greatest amount of water. On average, 70 percent of water consumed by humans is used to grow food. The proportion is even greater in regions with extensive agricultural activities such as California and Florida. The amount of water required to grow an agricultural product depends on a number of factors, including the local environment (particularly temperature and humidity), the method of irrigation, and the type of crop being grown. Some methods, such as the center pivot irrigation used in many southern states of the U.S., lose a large amount of water to evaporation before it reaches the plants. Drip irrigation, which delivers water directly to the roots of plants, requires less water. However, some crops require more water than others, even under identical conditions. The wide difference in the water requirements of various crops is shown in Table 1. Among the highest water consumers are bananas, whose trees produce a crop only once, and alfalfa, grown primarily as feed for cattle.

The range of water requirement for each crop, indicated by the +/− sign, represents variations in the amount of water needed under different growing conditions. Producing 1 kilogram of finished rice requires 2,500 to 3,400 liters of water and the production of 1 kilogram of refined sugar uses 1,500 to 2,000 liters of water. Growing alfalfa, an important source of protein in the feed of farm animals, requires almost twice the amount of water as most grains. Alfalfa grown in the United States is used as feed in the dairy industries of many countries, including China, Japan, the UAE, and Saudi Arabia. The demand for alfalfa in the international market increases the price of dairy products in the United States.

Growing cotton also requires copious amounts of water. Nearly half the clothes and other textiles in the world are made of cotton, while much of the rest are made from synthetic fibers that are petrochemical products. It takes more than 20,000 liters of water to produce the 1 kilogram of cotton that would produce a single T-shirt and a pair of jeans. Because of its large water requirement, 73 percent of the global cotton harvest comes from irrigated farmlands. Since cotton plants are highly susceptible to insects and disease, farms that grow cotton use 24 percent of the insecticides and 11 percent of the pesticides sold in the world, even though cotton is farmed on only 2.4 percent

Table 1 Water Required By Seasonal Crops.[107]

Crop	Water Need (mm/total growing period)
Banana	1700 +/– 500
Alfalfa	1200 +/– 400
Citrus	1050 +/– 150
Cotton	1000 +/– 300
Sunflower	800 +/– 200
Pepper	750 +/– 150
Maize (corn)	650 +/– 150
Sugar-beet	650 +/– 100
Tomato	600 +/– 200
Peanut	600 +/– 100
Potato	600 +/– 100
Rice (paddy)	575 +/– 125
Barley	550 +/– 100
Millet, Sorghum	550 +/– 100
Oats, Barley	550 +/– 100
Sorghum	550 +/– 100
Wheat	550 +/– 100
Soybean	525 +/– 75
Melon	500 +/– 100
Onion	450 +/– 100
Cabbage	425 +/– 75
Pea	425 +/– 75
Beans (Legumes)	400 +/– 100

Source: FAO.

of farmland. The use of chemicals and large demand for water greatly increases the burden of cotton farms on the environment.

The production of sugar, either from sugarcane or sugar beet, also consumes large quantities of water. Sugar is a product that most of us consume daily. Over 145 million tons of sugar is produced in 121 countries each year, occupying a large proportion of farmlands in the world. About two-thirds of sugar is made from sugarcane and the rest from sugar beets. Heavy sugar cultivation over the last few decades has substantially decreased the productivity of farmland in many parts of the world. With the exhaustion of nutrients in the soil and pollution of land, additional farmland is frequently added to sugar plantations by clearing the land of forests and other vegetation, destroying the habitat of many forms of plant and animal life. According to the World Wildlife Fund, sugar may be responsible for more biodiversity loss and damage to the ecosystem than any other single crop. In addition to the destruction of the habitat of numerous forms of life, sugar plantations require large amount of water for irrigation, and numerous agricultural chemicals in the growing period.[108] In the United States, sugarcane plantations on tens of thousands of acres in Florida have seriously degraded the unique ecosystem of Florida's Everglades. This area used to be a tropical forest supporting innumerable life forms and has now become lifeless marshland from the runoff of fertilizers.

Rice, cultivated in almost all countries of the world, is another crop that requires large amounts of water. It takes about 2,291 meters^3 of water to produce one ton of paddy rice, against 1,334 meters^3 of water per ton of wheat. The discrepancy is even greater because husked rice—that a consumer buys—requires 3,420 meters^3 of water per ton due to the loss of weight in the process of removing the husk. Since rice is the common staple in many countries, the amount of rice grown in the world is about the same as wheat. The total amount of water used to grow rice is almost 2.5 times the water used by wheat farms. Rice cultivation also adds to the burden of greenhouse gases in the atmosphere because fields are flooded with water in the early stage of the growth of rice plants. At this stage, many greenhouse gases are emitted, including carbon dioxide, methane, and oxides of nitrogen. Higher temperatures caused by global warming increase the emission of the potent greenhouse gas methane, thus further exacerbating the process.[109] The International Rice Research Institute is now developing better management options to make rice farming more productive, eco-friendly, and resilient to climatic extremes and other challenges.

Agriculture in many regions of developing countries is primarily rain fed, although it is supplemented with other sources of water wherever possible. The output of farms that depend only on rainfall is significantly lower than those that are irrigated. Some farms are irrigated by diverting the water of rivers with canals to farmlands. However, this can only be done where water in the river is plentiful and the fields are within a reasonable distance from the river.

The alternative, widely practiced these days, is to dig for groundwater near or underneath the agricultural fields.

Industry also uses a substantial amount of water which, in many places, is second only to the water used for agriculture. Almost every industry uses water in some phase of operation, including processing, washing, cooling, and transporting raw material to factories and the final product to distribution outlets. Industries that are involved in processing food items, and those that produce paper and chemicals, require more water than others. Among all industries, electric power plants stand out as the greatest users of water for cooling, whether they use fossil fuels or nuclear power as a source of energy, creating a nexus between power production and water consumption. Power generation requires large amounts of water in various stages of operations, although some of it is returned back to the source. Water is used in drilling and mining of natural gas, coal, oil, and uranium. Refining these products before they can be used as fuels also requires substantial amounts of water. The emission of pollutants like sulfur and mercury from thermoelectric plants is minimized by circulating water in smokestacks. The cooling systems of power plants use water in both once-through and recirculating systems. In once-through systems, water is continuously withdrawn at a rapid rate and returned to the source. The water returned to the source in these systems is at a higher temperature and often polluted with rust and other materials. The higher temperature of the returned water is injurious for aquatic organisms. Water withdrawn in recirculating systems, comparatively smaller in amount, is essentially lost due to evaporation.

On average, coal and nuclear power plants use 20 to 60 gallons of water for each kilowatt-hour of electricity produced. Electric power generation was responsible for more than 40 percent of freshwater withdrawal in the United States in 2008—about 100 billion gallons per day.[110] Various phases of electric power production also affect water quality because minerals unearthed during fuel mining and drilling can contaminate groundwater. Combustion of coal creates waste with dangerous toxins such as mercury, lead, and arsenic. Some of these chemicals eventually end up in the groundwater.

Groundwater

Groundwater is the largest source of usable freshwater in the world. Most of it consists of water from rainfall that has been stored in geologic formations of soil, sand, and fractured rocks beneath the Earth's surface. Groundwater is an important source of fresh water where other sources are not available. Water stored in aquifers can be extracted by drilling with pumps or wells. In the United States, about 120 million people living in rural areas depend on underground aquifers for daily necessities. In addition, it is also used to irrigate crops and in the manufacturing process. Extracting more water from the aquifer than the rate of recharge depletes it and lowers the level of groundwater; there is a limit to which this can be done before the wells run dry. Rain-fed aquifers in many parts of the world are being depleted at a rapid rate. While the demand for water is increasing, greater uncertainties in rainfall due to climate change and paving land with concrete for buildings, parking lots, etc. reduces the amount of water that can be absorbed by the ground.

The three major grain-producing countries are the United States, China, and India. There is an acute shortage of water in many parts of India and China. In the ultimate analysis, the productivity of farmlands is limited by the amount of available water and, as the sources of water dry up, those lands will not be able to grow sufficient crops. Using data collected by NASA's satellites from 2003 to 2013, researchers found that most major rain-fed aquifers in the world are depleting faster than the rate of discharge. It was found that twenty-one of the world's thirty-seven largest aquifers have passed the tipping point from which recovery will be extremely difficult.[111] Because of excessive withdrawal, the level of groundwater in many cities of the United States, including Houston, Memphis, Louisiana, Mississippi, and Tennessee, have declined substantially.[112] Depletion of aquifers is a global phenomenon, affecting people in many regions of the world.

Fossil Aquifers

Several aquifers that stored water eons ago, when the local climate was very different from today's dry climate, are known as "fossil aquifers." Thousands or even millions of years ago, rainfall or runoff was plentiful in these regions and water accumulated in stable rock formations. Since these aquifers are not recharged now, they represent a precious and exhaustible resource. Once they are pumped dry, the water is gone forever. Fossil aquifers include the Nubian Sandstone Aquifer System beneath the Sahara Desert, the Arabian Aquifer, the Ogallala Aquifer under the great plains of the United States, and the aquifer in the North China Plains. These aquifers are located in regions that have very little rainfall these days and agriculture is sustained by pumping water from them. Although the amount of water in these aquifers was once very large, rapid depletion during the last few decades has substantially decreased the stored volume of water in all of them.

The Ogallala Aquifer, also known as the High Plains Aquifer, is the largest fossil aquifer in the world. It supplies water to most of the irrigated farmlands in the Central United States. It covers an area of 175,000 square miles (450,000 square kilometers) in parts of eight states: South Dakota, Nebraska, Wyoming, Colorado, Kansas, Oklahoma, New Mexico, and Texas. Water accumulated in this aquifer over 20 million years ago when gravel and sand from the Rocky Mountains were eroded by rain and washed downstream. Those sediments soaked up water from rain and melted snow and preserved it in the aquifer for millions of years in sponge-like formations. Some water from Ogallala provides for the needs of local communities for household activities and for industrial operations, but the bulk of it is used to irrigate the farms that produce a large portion of agricultural items in the country. It has fostered the development of industrial agriculture and redefined the landscape of Central United States by promoting economic expansion and population growth. The Ogallala Aquifer supplies most of the water for growing the corn that is required by the industrial animal farming facilities. It has been estimated that more water is withdrawn from Ogallala every year to grow feed for cattle than is used to grow fruits and vegetables in the entire country.

Farmers in the aforementioned eight states use powerful, multi-stage pumps to extract water from Ogallala at a rapid rate, and use this water to irrigate fields of corn, wheat, and alfalfa. Since these states are in arid regions, the rate of recharge of the aquifer is minimal, and it is losing its reserve of water. A study led by David Steward of Kansas found that 30 percent of the Kansas portion of Ogallala Aquifer has already been pumped out, and another 39 percent will be used up in the next few decades at existing pumping rates. As a result, agricultural production is likely to peak around 2040, and then begin to decline.[113] The U.S. Department of Agriculture (USDA) reports that in parts of Texas, Oklahoma, and Kansas—three leading grain-producing states—the underground water table has dropped to more than 30 meters (100 feet) below the surface.

Before the Ogallala Aquifer was discovered, much of the land in the Southwestern United States was considered suitable only for grazing cattle and not for intensive farming. Water from the Ogallala Aquifer transformed the land into one of the most productive regions of the world and redefined the landscape of this large area. Without water from this aquifer, vast stretches of land would revert to their original desert-like state. Since the United States is a major exporter of agricultural products, this change would have a serious impact on the availability of food everywhere. It will also limit the production of meat because 70 percent of the grain produced in the country is fed to livestock. Globally, about 80 percent of soy and 40 percent of corn produced in the world is fed to livestock and poultry.

The Nubian Sandstone Aquifer System provides precious water to Libya, Chad, Sudan, and neighboring countries. Libya is primarily a desert—much of its habitation depends on water from the aquifer under the Sahara's sands, discovered during oil explorations in the 1950s. About four decades ago, the Saudi government realized that the Arabian Aquifer contains a large amount of water (120 cubic miles, or 500 cubic kilometers). It subsequently embarked on a major project to irrigate its fields with water from this fossil aquifer with the eventual goal of becoming self-sufficient, and even an exporter of wheat. By 2008, however, it was discovered that intensive farming had used up 80 percent of the water in the aquifer.[114] A giant freshwater resource in the hot and dry Saudi desert was severely depleted in one generation. The government abandoned the plan to grow food locally and Saudi Arabia now imports all grains needed to support its population. Since the fossil aquifers in China and Saudi Arabia are being emptied rapidly, the water level at some places has fallen so much that the wells have to be dug to a depth of 1,000 meters (0.62 miles).

Another problem with extracting large quantities of groundwater is that it may cause the land in some regions to collapse. Water is needed to maintain the integrity of the underground layers of rocks and sand. When water is extracted, the land may sink, either gradually or all of a sudden. This phenomenon is known as subsidence and it is happening in many parts of the world, including the United States.[115] The problem of sinking ground is most acute in California, where groundwater supplies about 60 percent of the state's water, most of which is used to support agriculture. According to the U.S. Geological Survey, the sinking of the land is starting to destroy bridges, crack irrigation canals, and twist highways across the state. Some places in the state are sinking more than a foot per year. It will take a huge amounts money to repair the damage that has been done; the problem has been compounded by the recent droughts in the state.

Water Pollution

While the scarcity of water is affecting many regions of the world, the related problem of pollution decreases the availability of usable water in numerous places. Pollution of water by chemicals or pathogens makes it unsuitable, even dangerous, for consumption by humans. It may also be harmful for other forms of life and is bad for the ecosystem. Water may be polluted by individuals, industries, or agriculture. Individuals pollute water by discharging household chemicals in the ground or in the sewage system. Households in most developed countries use a plethora of chemicals for cleaning and washing, and to kill pests and microorganisms. In addition to posing an immediate risk to people exposed to them, most of these products eventually end up in the groundwater.

The discharge of untreated municipal waste in bodies of water contaminates them with harmful chemicals and pathogens. Water with significant organic matter may lead to the growth of waterborne bacteria such as those that cause cholera, typhoid fever, and bacillary dysentery. The most common effect of drinking water contaminated with pathogens is diarrhea. It is estimated that dehydration caused by diarrhea results in the death of 1.5 million people each year, mostly young children in developing countries. Other diseases that may be caused by waterborne pathogens include polio and hepatitis.

Most factories use chemicals that are harmful for human health. Chemicals escaping from these facilities pollute the local environment—air, water, and land. Water is polluted when factories discharge industrial effluents in rivers, streams, and ground. While some discharge of effluents is allowed by the municipalities for the economic benefits brought to the community by the manufacturing facilities, the pollution from these facilities often exceeds these limits. Factories that produce sugar, for example, routinely discharge the polluted wastewater in neighboring rivers or streams in violation of local regulations. In addition, accidents in factories that release large amount of dangerous chemicals in the ground often occur in all parts of the world.

Coal-fired power plants, which produce about 30 percent of the electricity in the U.S., are the biggest polluters of water. The strip-mining method of extracting coal contaminates local bodies of water. Additionally, coal is usually washed to remove sulfur and other impurities to decrease the emission of harmful gases and vapors from the smokestack of power plants, but this process results in pollution of the water of rivers and streams. The gaseous emissions from the power plants usually contain sulfur dioxide, nitrous oxide, and mercury, which mix with rain or snow to pollute the water of lakes, rivers, and groundwater. Power plants that operate with coal, oil, or natural gas consume large amount of water because they use steam to drive the turbines.

The quality of drinking water in the U.S. was improved with the passage of the Safe Water Act in 1974 and subsequent actions by the government. However, millions of people in the country are still exposed to polluted drinking water. The ongoing risk of polluted water was highlighted a few years ago by the plight of residents in Flint, Michigan. The Clean Water Rule of 2015, passed during the Obama administration, gave the federal government the authority to protect waterways from pollution. However, the Trump administration decided to repeal this rule and replace it with an industry-friendly rule that would cover fewer waterways. This change was opposed by environmental groups.

Water Pollution from Industrial Farms

Conventional industrial farms use chemicals in all phases of operations. Synthetic fertilizers containing nitrates and phosphates are applied to promote the growth of plants, and herbicides and insecticides are used to kill weeds and insects. Plants absorb only a fraction of the fertilizer applied to the crops through their roots, the remainder of the fertilizer and most of the pesticides migrate to groundwater and contaminate neighboring bodies of water such as lakes, ponds, and rivers, eventually reaching the coastal seas. Once groundwater is contaminated with harmful pollutants, it is exceedingly difficult to bring it back to a purified state. The variety of pesticides used by farmers and gardeners is very large, many of which are toxic and dangerous to humans and aquatic life. Some of these chemicals can affect human health by causing neurological and reproductive disorders while others may be carcinogens.[116] Some pesticides are suspected to disrupt the hormone messaging system of humans and wildlife. Studies have shown that chemical pesticides persist in the ground and waterways long after their application on farms, resulting in the accumulation of these harmful chemicals in the environment. Pesticides do not just kill target pests; they also target beneficial insects in and around the fields to which they are applied.

Water-soluble nitrates from fertilizers that are not absorbed by the roots of plants contaminate drinking water. Infants who drink water high in nitrates can become seriously ill with shortness of breath and blue-tinted skin, a condition known as "blue baby syndrome" that can be fatal. Due to the extensive use of synthetic fertilizers, nitrate levels in water have been increasing in many regions of the world. Excessive nitrogen in drinking water can interfere with the transport of oxygen in blood, causing an acute sickness. Nitrates and phosphate in water that flows from farmlands to coastal seas cause an explosive growth of algae known as an algal bloom, which leads to the formation of dead zones devoid of all forms of aquatic life. There are hundreds of dead zones in coastal seas all over the world. According to the USDA, nitrogen fertilizers applied to farmlands in the Midwest flow down the Mississippi River, creating a dead zone in the Gulf of Mexico that covers an area of 8,000 square miles and eliminates all forms of marine life from this highly productive portion of the sea.

Corn fields in the U.S. generally use atrazine to eliminate unwanted vegetation. This herbicide is a known hormone disruptor and is thought to increase the risk of miscarriage, reduce male fertility, and increase the chance of birth defects. The Cancer Panel appointed by the Obama administration classified atrazine as a possible carcinogen.[117] The U.S. Geological Survey found that 90 percent of the samples taken from a shallow groundwater portion of Ogallala contained elevated levels of nitrates and 14 percent contained atrazine. The runoff of fertilizers and pesticides is slowly accumulating in the groundwater in many regions of the world.

Although cotton is considered to be a natural product, its plantations depend on the application of large amounts of chemicals on a regular basis. In the United States, 84 million pounds of pesticides and 2 billion pounds of fertilizers were applied to the 14.4 million acres of cotton farms in 2015. Seven of the fifteen pesticides commonly used in cotton farms in the United States are classified as known or likely human carcinogens. Cotton farms usually employ defoliants on the mature plants as harvesting aids to open the green bolls. These defoliants are the most toxic farm chemicals currently in the market.

The ecosystem consists of plants, animals, bacteria, and fungi interacting with each other in complex ways to maintain a dynamic balance. The interactions include producer, consumer, predator, and prey relationships, and there is an exchange of water and oxygen between species. Pollutants in water may be fatal to some of these organisms, with an effect that may not be visible to the naked eye but may disturb the ecological balance, adversely affecting the growth and survival of species far removed from the one originally eliminated by the pollutant. An oil spill in the ocean may appear to kill only some large fish and birds but it also reduces the number of microscopic phytoplankton that are at the base of the marine food web. Although it may appear that the effect of pollutants is limited to a small region, a huge degradation of air or water has some effect on the entire biosphere. At places far removed from the source of pollution, the effect of the pollutants will be very small. However, as events of this type continue at an ever-increasing rate, small effects will add up and significantly degrade the environment.

Water Pollution from Industrial Livestock Facilities

Most meat and dairy products are produced in industrial facilities known as Concentrated Animal Feeding Operations (CAFO). These outfits are enormous—a single CAFO may contain thousands of large animals such as beef cows or tens of thousands or even hundreds of thousands of small animals. The waste of these animals is collected in large ponds called lagoons. These facilities make extensive use of antibiotics, hormones, and pesticides, which get mixed up with the waste of animals in the lagoons. The biological content of these lagoons promotes the growth of pathogens of all kinds, some of which may even be resistant to the antibiotics used in the facility.[118] The typical method of disposing of the toxic slurry from the lagoons is to spray it on farms. Unfortunately, the volume of the slurry from a CAFO is so large that it cannot be absorbed by neighboring farmlands. The excess slurry pollutes and contaminates the groundwater with dangerous chemicals and pathogens. If the contents of lagoons are applied to farms when fruits or vegetables are ready for harvest, they become contaminated with dangerous bacteria such as E. coli. Water pollution from the manure can lead to serious environmental damage and harm to human health. Animal products from factory farms use more resources—particularly water—and pollute more than animal products from grazing systems.

According to the Centers for Disease Control and Prevention (CDC), pathogens likely originating from factory farms have been responsible for numerous outbreaks of diseases in the country. A spill from just one lagoon can pollute a whole region. In August 2005, 3 million gallons of animal waste poured into the Black River in New York State, polluting an exceptionally large area and killing around 250,000 fish in the river. It is estimated that there are 9,900 CAFOs in the U.S. and such facilities are becoming popular in many other countries.

Water Pollution from Home Lawns

In the United States, home lawns and turf grass cover about 40 million acres, an area three times larger than that of irrigated corn. Turf grass requires a plethora of chemicals on a regular basis. The ground is cleared of unwanted vegetation by applying an herbicide such as glyphosate (Roundup). Fertilizers, which consist primarily of nitrogen in the form of urea or nitrates, are also used extensively. In "slow release" fertilizers, ammonia is mixed with urea and formaldehyde or is encased in sulfur or a synthetic polymer. Most attractive lawns are not made of local grasses and require extensive care during the growing season. Turf grass requires regular application of herbicides to control weeds before they emerge and also to kill broadleaf weeds, crabgrass, and other invasive plants. Another set of chemicals is required to control insects that may injure the turf. It has been estimated that 3 million tons of these chemicals are applied on home lawns each year. Many of these chemicals are harmful for human health—some of them are endocrine disruptors while others may cause birth defects, liver and kidney damage, and cancer. Scientific studies have found that dangerous chemicals, such as the weed killer 2,4-D and the insecticide carbaryl, contaminate air and get deposited on carpets and surfaces inside homes with foot traffic, thus exposing residents, including children, to these chemicals.[119] Turf grass requires more water than many agricultural crops. A substantial portion of chemicals applied to home lawns ends up polluting the groundwater. Overall, home lawns use a lot of water and cause substantial pollution of water and land. Since a shortage of freshwater is predicted in many parts of the country, home lawns have become a harmful luxury.

Biofuels

Biofuels consist of ethanol or biodiesel produced from grains, sugar, and other organic materials. In recent years, many countries have begun pursuing this source of energy to reduce their dependence on fossil fuels. In addition, some studies have shown that the combustion of biofuels produces less greenhouse gases than fossil fuels. For these reasons, governments in the U.S. and some other countries have mandated that ethanol made from plant sources should be nixed with gasoline. Major producers of biofuels are Brazil, the United States, China, and India. In 2000, less than 5 percent of the corn crop in the U.S. was used to produce ethanol. However, the proportion has now increased to 40 percent, producing 13 billion gallons of ethanol.[120]

A comparison of the environmental impacts of biofuels and fossil fuels must consider the whole production cycle of these two sources of energy. In many cases, land has to be cleared of vegetation to grow the required crops. In Brazil, this is done by razing rainforests on a large scale. The loss of forests removes an important sink of carbon dioxide, thereby increasing the concentration of greenhouse gases in the atmosphere. Biofuel crops such as corn and sugarcane require a lot of water with the result that it takes, on average, 3,320 liters of water to produce one liter of biofuel.[121] Since there is already a shortage of water in many countries, this is an important consideration. The amount of water in the aquifers in the U.S. is large but limited, and there are already indications that the water table is falling in many areas. It is estimated that the ethanol plants located in the Southwest withdraw 15 million gallons of water per day. Withdrawal of such a large amount of water puts an additional pressure on an already stressed source of groundwater. The production of corn ethanol pollutes the land with chemicals, causes soil erosion, emits greenhouse gases, and reduces the availability of food. Since ethanol contains only two-thirds the energy of pure gasoline when burned, mixing ethanol in gasoline reduces the efficiency of vehicles by 3 to 4 percent.

Crops required to make biofuels are grown in monoculture, i.e., a single crop is grown over a large area. Fertilizers and pesticides have to be applied to these crops on a regular basis, leading to pollution of local bodies of water. Diverting crops to make biofuels increases the cost of food in local and international markets. The federal government in the U.S. provides an array of tax breaks and subsidies to increase the use of corn ethanol. The area of farms growing corn increased by more than 13 million acres from 2006 to 2013. Most of the additional farmland was obtained by replacing the crops of other grains, such as wheat, oats, sorghum, and barley. In all, U.S. corn producers received subsidies totaling $90 billion between 1995 and 2010.[122] These subsidies have created a powerful corn lobby that opposes any changes in the Renewable Fuel Standard, which mandates that gas and diesel fuels be mixed with ethanol. The ultimate cost of these subsidies is passed on to American consumers in the form of higher gas prices and increased costs of many food items.[123] When all these factors are taken into consideration, the production of biofuels may not be advantageous at all.

Virtual Water

The projected shortage of water in many parts of the world makes it important to consider how and where it is consumed. Virtual water is defined as the volume of water required to produce a commodity. Each agricultural product has virtual water associated with it, which is the amount of water required to produce it. All commodities—not only agricultural products—have virtual or embedded water content, although it is significantly less for non-agricultural items. Arjen Hoekstra, professor of Water Management at the University of Twente, the Netherlands, and his colleagues have done seminal work in calculating the water contents of various food items and also the virtual water traded between countries. The concept of virtual water is useful when the amount of water available is limited because it can help in developing an integrated approach to water management.

The amount of water required to grow seasonal crops was given in Table 1. The amount of virtual or embedded water contained in some familiar items in units that are used in households is given in Table 2.[124]

The average water consumption of a meat-eating person in the United States is about 5,000 liters per day, while that of a vegetarian is roughly half this amount.

Table 2 Embedded Water in Typical Household Items.

Item	Embedded Water
Pint of beer (568 ml)	170 liters
Glass of milk (200 ml)	200 liters
Cup of coffee (125 ml)	140 liters
Cup of tea (250 ml)	35 liters
Glass of orange juice (200 ml)	170 liters
Glass of apple juice (200 ml)	190 liters
Glass of wine (125 ml)	120 liters
One orange (100 g)	50 liters
One tomato (70 g)	13 liters
One potato (100 g)	25 liters
Bag of potato chips (200 g)	185 liters
One egg (40 g)	40 liters

(Continued)

Table 2 Embedded Water in Typical Household Items. *(Continued)*

Item	Embedded Water
Slice of bread (30 g)	40 liters
Bread with cheese (30 g + 10 g)	90 liters
Hamburger (150 g)	2,400 liters
Cotton T-shirt (250 g)	2,000 liters
Cotton pair of jeans	8,000 liters
Bovine leather shoe	8,000 liters
Chicken meat (1 kg)	3,900 liters
Pork (1 kg)	4,900 liters
Beef (1 kg)	15,500 liters
Cheese (1 kg)	3,180 liters

Note: 1 gallon = 3.79 liters.

Water Footprint

The concept of water footprint was developed by Arjen Hoekstra and colleagues as a measure of water use in a country.[125] The water footprint of a person, family, community, or country is the amount of water used for direct consumption, as well as the embedded water in food and all consumer products. This concept is somewhat similar to ecological footprint developed by Wackernagel and Rees, which represents the area of productive land and aquatic ecosystems required to produce an item and to assimilate the waste produced.[126] Calculations of water footprints at the national level take into account domestic water production and consumption, to which the embedded water in imports is added and from which the embedded water in exports is subtracted. The average global of water footprint is 1,240 cubic meters per person per year. The per capita water footprints of some countries are given below.[127]

The factors that determine the water footprint of a country are lifestyle, climate, amount of animal-based foods consumed by its people, and agricultural practices. The per capita water footprint of a few countries is shown in Table 3. The water footprint of people in the United States is the highest in the world primarily because Americans eat more animal-based foods—meat and dairy—than people in other countries, almost three times the world's average. Other contributing factors include lifestyle-related commodities, such as private swimming pools, and home lawns that must be watered on a regular basis during the

Table 3 Per Capita Water Footprints.[128]

Country	Average per Capita Water Footprint (gallons/day)
U.S.	1,800
Italy	1,670
Thailand	1,590
Nigeria	1,410
Mexico	1,010
Brazil	980
Indonesia	920
Pakistan	870
Japan	830
India	710
China	510

growing season. A high consumption of industrial products also contributes to water consumption. The water footprint of developing countries is considerably smaller because they usually do not have running water and modern amenities. However, evaporation due to higher ambient temperatures and poor agricultural practices increases the consumption of water. These factors partially explain the large water footprint of countries close to the equator, such as Nigeria.

Water Footprint of Animal-Based Foods

With increasing affluence, the consumption of animal-based foods (meat and dairy products) is increasing in many countries. The total annual meat consumption in China, for example, increased from 8 million tons in 1978 to 71 million tons in 2011 and is still increasing.[129] The water footprint of any animal-based food is larger than the water footprint of plant products of equivalent nutritional value by a substantial amount.[130] Among animal products, beef requires the greatest amount of water. Raising a cow for three years and slaughtering it to produce 200 kilograms of beef requires more than 3 million liters of water, amounting to about 15,400 liters of water per kilogram of beef. Although the water required per kilogram is less for other animal products, it is still much more than for cereals, fruits, and vegetables.

International Trade in Virtual Water

International trade represents a very large transfer of virtual water between countries; it allows some countries to both preserve internal water resources and consume water of countries that export items with large virtual water content. Since water requirements of crops differ by a substantial amount depending on the local climate and growing conditions, international trade in virtual water may be beneficial for the global water resource if crops are grown where conditions require less water and exported to countries where conditions are not so favorable. However, financial and political considerations play an important part in this process. The countries that export virtual water may be doing so without full consideration of the true value of the traded items or the long-term implications of the trade for its own inhabitants. Since the amount of fresh water in the world will soon be—perhaps already is—insufficient to meet the needs of all people, countries with enough financial resources will be able to buy virtual water from other countries that are in need of foreign currency or that respond to pressures and inducements of other kinds. Hence conservation of water will not be a priority for countries that have enough financial resources to import virtual water.

The total global fresh water used each year in the world is estimated to be 7,541 billion cubic meters (BCM). A substantial part of this—around 1,625 BCM—is traded across national borders. Agricultural products account for the largest share—roughly 61 percent—of the traded water, and about 17 percent virtual water is traded as livestock products. The rest is embedded in manufactured items. In general, countries with a fair amount of capital at their disposal tend to import agricultural products with large virtual water content from less developed countries, thereby saving domestic water to provide amenities to their inhabitants. International trade transfers the burden of water consumption to other countries. Even though the United States imports a substantial amount of virtual water in the form of agricultural commodities, it is the largest net exporter of virtual water in the world, exporting about 182.3 BCM of water. This amount is larger than the total water export by the next three countries—Australia, Canada, and Argentina. Other major exporters of virtual water are Thailand, India, France, New Zealand, Vietnam, and Brazil. The largest importers of virtual water are Japan, Italy, South Korea, the Netherlands, Indonesia, and China.[131]

Rapid economic growth, urbanization, and changing food preferences have greatly increased the requirement of agricultural products in many countries.

North America and South America are the greatest providers of virtual water to the world, and Western Europe, Central and South Asia, and the Middle East are the largest importers. The United States provides a substantial amount of virtual water to the Far Eastern countries (Japan, China, and South Korea), and to Mexico. The water footprint of European countries is large because they import many agricultural and livestock items from countries around the world. The requirement of agricultural products containing large virtual water content is increasing rapidly in China, which now relies heavily on imported virtual water to sustain its growth. China is now the biggest soybean buyer in the world, importing more than 80 million metric tons per year, mainly from Brazil and the United States. Each ton of soybean contains 2,000 cubic meters of embedded water. China and Japan are the world's largest grain importers. U.S. agricultural exports to China have grown more than 200 percent over the past decade. U.S. farm and food exports to China totaled more than $20.2 billion in 2015.[132]

Hay, consisting primarily of the water-thirsty alfalfa, is also very much in demand. Because of its protein content, alfalfa is used as feed for livestock. China imported a record 1.13 million tons of alfalfa and hay in 2015 from the U.S.[133] Ships that bring the latest iPhones, flat-screen TVs, household goods, and other gadgets from China for the U.S. market offer extremely low rates to take dried alfalfa bales back home. Because of its importance to the dairy industry, the export of alfalfa to China, South Korea, Japan, the United Arab Emirates, and Saudi Arabia from the United States has been increasing rapidly in recent years. From 2004 to 2017, the U.S. shipped an average of 2.8 million tons of hay overseas each year. The export of alfalfa to China reflects a larger trend in the U.S. international trade: America increasingly exports raw materials that China converts into valuable products—cotton into shirts, hides into shoes, wooden logs into furniture.

Land and Water Grab in the U.S.

Land grabbing, defined as the ownership or use of land by foreign investors or governments, is happening at an accelerated pace in many parts of the world. It not only involves the control of land but, more importantly, of freshwater resources, both green water (rainwater) and blue water (water from rivers, streams, and groundwater). Availability of fresh water is an important component of the acquisition of land in other countries; land without the availability of fresh water will not elicit any interest. Although many countries are active in land grabbing, China and the oil-rich Middle Eastern countries are particularly active in this enterprise. Such acquisitions bring minimal benefits in terms of jobs and financial support to the communities whose land and water have been appropriated.

The largest dairy company of Saudi Arabia, Almarai, bought 15 square miles of farmland in Vicksburg, Arizona, to grow water-guzzling alfalfa to be shipped back to its home country. Each of the fifteen water wells on the property can pump about 1.5 billion gallons of water from the underground Ogallala aquifer per year. The yield of hay from farms in Arizona is very high because hay has a very long growing season. The heat of the desert dries it quickly, making it easy to compact for shipping out of the country. However, high temperatures greatly increase the water required to grow alfalfa. Saudi-owned food giant Almarai also purchased 1,790 acres of farmland in Blythe, California. Blythe is an agricultural area that borders the Colorado River and hence has access to the river's water. The Saudi purchases are in areas where there are few restrictions on the use of water. An Emirati company, Al Dahra, has essentially replicated this land grabbing in Arizona and California, depleting the local groundwater to grow hay to support the UAE's farm-animal industry. Several other countries are beginning to grab land as well. In November 2012, AgriVest, a unit of Swiss banking giant UBS, purchased 9,800 acres of farmland in Southwest Wisconsin.

Acquisition of farmlands and the associated fresh water by agencies of other countries is a new phenomenon. The current scramble for land represents a worldwide struggle for control of fresh water. Although farmers do not care if their produce goes to China or Saudi Arabia, such operations are mortgaging our future and causing a loss of fresh water crucial for the productivity of the land. Due to extensive mechanization in these farms, any increase in employment in the communities is minimal. Laws in these regions were designed for local or domestic farming, with the acquisition of land by foreign powers, farmers become tenants on the property they once owned. Resentment over such deals is slowly growing. The idea that another country will become a major user of local fresh water did not exist a few decades ago, and there are no laws to prohibit such practices in the interest of the future of local economies.

There is a big difference in the process of growing crops by foreign corporations and by local farmers. Local farmers are sensitive to the available resources, including water and agricultural land, because their fate is tied to these things. They may grow fewer water-intensive crops if the water table is falling rapidly or take some other preventive actions. They will also avoid practices that permanently despoil the land with a buildup of salts or an application of inordinate amounts of chemicals that may adversely affect the land or the groundwater. The farmlands in Arizona and California that have been purchased by foreign corporations were originally used to grow a variety of crops but are now used exclusively to grow alfalfa that is shipped out of the country. These farms now consume much more water because alfalfa is a particularly thirsty crop. Foreign investors have no qualms about overexploiting local resources, since their only objective is to get the maximum benefit out of the land in a very short time. If the land runs out of resources or is degraded, they will simply move on to another site.

Grabbing land and water in other countries is now an international phenomenon in which countries with large financial resources appropriate the water of other countries. Worldwide, it is estimated that 500 million acres (200 million hectares), an area three times the size of Texas, has been sold, leased, or claimed by foreign agencies.[134] This trend is expected to speed up as water shortages become more evident. Such land deals are made possible with the assistance of local or national leaders. The general population is rarely consulted and is usually unaware of the full implication of such deals. Laws to prevent such appropriation of local resources by foreign agencies are either nonexistent or ignored. The countries that are most active in seeking land and resources elsewhere include Middle Eastern countries, China, and some northern European countries. The targeted countries include fertile countries such as Brazil, Russia, and Ukraine, but also poor countries like Ethiopia, Cameroon, Madagascar, and Zambia. According to Devinder Sharma, analyst for the Forum for Biotechnology and Food Security in India: "The environmental tab of highly intensive farming—devastated soils, dry aquifer, and ruined ecology from chemical infestation—will be left for the host country to pick up."[135]

In an extension of the trend of acquiring resources in other countries to satisfy domestic need, Shuanghui International, China's largest pork producer, struck a deal to purchase Smithfield, the largest pork producer in the U.S., for $7.1 billion in 2013. Thus, the hogs raised in the modern and efficient pig farms and processing facilities of Smithfield will be exported to China. Based on official Chinese data, two-thirds of the waterways in that country are polluted and 10 percent of farmlands are contaminated with heavy metals. The acquisition of Smithfield gives China access to America's safe farmland and clean water supplies. Additionally, while the pork that is exported out of the country uses local resources, the waste produced by the animals remains in the U.S., degrading the local environment.

International Water Wars

Present and projected risk of water shortages is giving rise to conflicts between nations that share a source of fresh water, which may be a river, stream, or groundwater. These "water wars" have already started in some parts of the world and are expected to intensify in the coming years as the shortage of water becomes more acute. There are several flashpoints in different parts of the world that may lead to widespread conflicts.

The Nile is the longest river in the world. The lakes and tributaries of this river provide water to nine African countries before it reaches the Mediterranean Sea. Egypt has used military force to ensure their control over the headwaters of the Nile because the country has no other source of water. Sudan, Ethiopia, and Uganda have constructed various river projects to increase their withdrawal of water from the Nile. The Ethiopian government is now in the process of constructing a 6,000-megawatt hydroelectric dam on the Nile, which will be the largest such project in Africa. Since the dam will likely decrease the flow of water in the river, there is strong opposition to its construction by neighboring countries.

The Brahmaputra River originates in Tibet and flows through India, ending up in Bangladesh. China is constructing several hydroelectric plants to harness its water. These projects may decrease the flow of water in the river to the detriment of Indian and Bangladeshi farmers, who depend on the water of this river for irrigation to produce crops in these fertile lands. A substantial reduction in the flow of the Brahmaputra would be devastating for millions of people. Although some progress has been made in resolving the dispute, the Brahmaputra River remains a potential source of friction between China and India.

In the Middle East, the Jordan River Basin is shared by several countries, including Lebanon, Syria, Israel, and Jordan. The Tigris and Euphrates rivers originate in Turkey and flow through Iraq and Syria; the scarcity of water in these regions has been a source of conflict for a very long time. Turkey is planning to construct twenty-two dams and nineteen hydroelectric plants in the

Tigris-Euphrates basin in the Southeastern Anatolian Project. This project will be extremely detrimental for Iraq and Syria, which have been experiencing droughts for a number of years. In another region, the Mountain Aquifer underneath the West Bank is a point of contention between Israel and Palestine.

Some conflicts originate not in the quantity of water but in its quality. Regions that have suffered through wars for extended periods have their air, water, and land polluted by leftover chemicals and weapons used in times of war. If the demand and need for water is greater than its availability in any region, there are bound to be disagreements that may lead to armed conflicts.

Control of Water by Corporations

There are disagreements between human rights groups and industrial houses on the availability of water to all people. The notion that water is a fundamental basic right is challenged by corporations that see an excellent opportunity in the projected shortages of water. Peter Brabeck-Letmathe, chairman of Nestlé—the largest manufacturer of food products in the world—sits on the board of many multinational corporations and groups of corporate CEOs. He thinks that water is being given at throwaway prices and that "reasonable pricing policies" will lead to more efficient use of water.[136] He says that privatization is the best way to ensure fair distribution and efficient use of this most important raw material. In this way of thinking, corporations should control all water on the planet, with the implication that it would be available only to those who can afford it.[137] In keeping with the belief that the ultimate social responsibility of the chairman of a corporation is to make as much profit as possible, Nestle extracted 36 million gallons of water from a national forest in California in 2015 to sell as bottled water. For this operation, Nestle paid only $524 a year to the San Bernardino National Forest.[138] During this time, Californians were ordered to cut their water use due to a historic drought in the state.

Privatization of water is advocated by many individuals and corporations. Since corporations are accountable only to shareholders and not to the public, privatization of water will lead to higher prices. Since the quality of water is not obvious to an average person, corporations may reduce costs by lowering water quality. Once a municipality signs over all or part of its water facility to a private water company, withdrawing from the agreement may be almost impossible because the company may invoke multinational trade agreements like NAFTA and GATT to obtain a ruling in its favor.

Water is the most basic human need, and access to water should be a fundamental human right. This concept is implicitly or explicitly supported by international law and the United Nations. Handing over water rights to corporations whose avowed objective is to maximize profits may endanger the lives of hundreds of millions—even billions—of people. International aid agencies and local communities should work to provide all humans with the basic water requirement.

The Way Forward

With the present and predicted shortage of water in many parts of the world, it is important that water footprints be reduced everywhere. The first step is to develop a consciousness of water consumption in daily activities. Future calamity caused by the shortage of water can only be averted if we realize the extent to which our collective lives depend on this precious resource and how close we are to reaching its limits. Changes in weather patterns brought about by global warming may affect the availability of water so that regions that have ample water right now may face a shortage in the coming decades.

Individuals can reduce their own water footprints and put pressure on industries to reduce the use of water. Governments should also be pressured to offer inducements for conserving water and penalties for waste. In addition to minimizing usage and waste, people should not pollute water with dangerous chemicals. In efforts to sanitize houses and get rid of pests, many people use chemicals of various kinds that are either toxic or harmful for human health and the environment in various ways. Home lawns consume a fair amount of water and also pollute the groundwater with fertilizers, insecticides, and weed killers. Water pollution should be considered as wastage because water is extremely difficult to purify. The polluted water has limited use and may even be harmful to human health.

Drastically reducing the consumption of meat is a necessary first step because the amount of water required by a meat-based diet is almost twice that of a plant-based diet. Numerous studies have shown that a diet of grains, fruits, nuts, and vegetables is good for the environment, and also for personal health. Clothes made of cotton are very popular because cotton has unique qualities that have not been replicated by synthetic materials. However, growing cotton requires a lot of water and the use of many chemicals that degrade the land and pollute the groundwater. Fortunately, there is a recent trend of growing organic cotton without the use of chemical fertilizers, pesticides, or defoliants. Right now, organic cotton only represents less than 1 percent of the cotton grown in the world. Several companies, including Patagonia, Nike, and Timberland, are now selling clothes made with organic fiber. The National Wildlife Federation is also launching a line of apparel made with organic cotton. The price difference between conventional cotton and organic cotton will decrease as its popularity increases. However, unlike synthetic fabrics, cotton cannot be recycled. Therefore, fabrics made of cotton should be used as long as possible and not discarded with the slightest change in fashion. Consumers should consider synthetic fabric to be a viable option.

On a nationwide basis, industries should adopt techniques that require less water. Ultimately this may happen anyway with the scarcity of water, but it is prudent to avoid reaching that state. The use of water in agriculture can be reduced by adopting techniques such as drip irrigation, which uses less water by delivering it directly to the roots of plants. Organic farming requires somewhat less water because of the capacity of humus in the soil to absorb and retain moisture for longer periods. Since the availability of water will, in the ultimate analysis, determine the amount of food that the planet can produce, it is imperative that all actions be taken to minimize the water footprint everywhere. Shortages of water will have a serious impact on human life and will threaten the very survival of some sections of people.

THREE
· · · · · · · · · · · · · · · · · · · ·

ENVIRONMENTAL DEGRADATION

Humans depend on a healthy biosphere for survival and well-being. In addition to the basic requirements of air, water, and food, many other ecosystem services are necessary for our welfare. Forests convert carbon dioxide into oxygen and also moderate the climate. Oceans provide us with food and maintain the balance of gases in the atmosphere. Insects pollinate the flowers of plants and trees, which provide us with fruits, nuts, and vegetables. Numerous microorganisms convert our waste into useful products. The ecosystem is vital to human existence, but it is being severely damaged in many ways. Environmental degradation is the reduction in the capacity of land and seas to meet the needs of humanity. It takes many forms, ranging from pollution and destruction of ecosystems to the degradation of arable lands and sources of fresh water.

The primary cause of environmental degradation is careless and excessive use of planetary resources beyond the regeneration limits. The continuously increasing demand for comforts, conveniences, and novel objects, aided by technological developments, has greatly increased the use of natural endowments in recent decades. The entire planet is searched for things that may provide us with food or energy, or that may be useful in some other way. From the deepest oceans to the highest mountains, there is nothing that is beyond the reach of humanity. We are using planetary assets and generating waste at a rate that cannot be sustained. According to an analysis by the research organization Global Footprint, we would need about 1.6 Earths to provide for the present lifestyle in a sustainable manner.[139] This overshoot is achieved by using the built-in reserves of the planet, and leaving degraded materials and waste for posterity. The Earth is not able to recover the loss of ecosystem assets like fresh water, marine life, forests, and topsoil that are consumed or degraded each year by human activities, sending the planetary resources into a downward spiral. The pace of exploitation of natural reserves is accelerating instead of slowing down. Since the stored resources of the planet are large but finite, the imbalance between the generation and consumption of natural resources will lead to a clash with dangerous consequences.

One way in which the burden of our lifestyle on the ecosystem is measured is by calculating the Earth Overshoot Day, the day on which humanity's resource consumption exceeds Earth's capacity to regenerate those things for that year.[140] This day arrives earlier and earlier each year—in 2019, it was July 29, which means that in seven months humanity used the resources generated by the planet in the entire year.[141] However, the economic slowdown caused by the COVID-19 pandemic pushed the Overshoot Day forward to August 22 in 2020. Although the Earth Overshoot Day is calculated at the global level, the exploitation of Earth's resources varies greatly among countries because planetary resources are used up at different rates depending on the lifestyle of the inhabitants. Each country's Overshoot Day has been calculated—it is the date when the planetary resource will be used up if people everywhere lived like the average person in the country under consideration. It indicates the relative use of Earth's resources by people in each country. The Overshoot Day occurs in February or March for most oil-producing countries, indicating that the planetary burden imposed by the lifestyle of the residents of those countries is very high. The United States is almost in the same league, with the Overshoot Day in mid-March. Most European countries and Japan have the Overshoot Day in May, ahead of the Global Overshoot Day, indicating that their lifestyle presents a greater burden on the planetary resources than the global average. The Overshoot Day was June 13 for China, July 13 for Brazil, and July 29 for India.[142]

In addition, there are many things not included in ecosystem services. The stock of many minerals that must be dug out of mines is decreasing. Some of these elements and compounds are essential for manufacturing electronic devices that are now commonly used everywhere, while others are important for agricultural and industrial activities.

The environment on which we depend for our survival and welfare has substantially degraded in the last few decades. It is useful to remember that the environment is a global resource and pollution does not recognize national boundaries. In today's globalized world, degradation in one region affects the entire world, and pollution in one place slowly permeates to distant lands through the slow process of migration and diffusion. Pollution and contamination take a toll on human lives and may ultimately determine the sustainability of our lifestyle. Being cognizant of the present state, and closely monitoring changes in it, is the essential first step for taking remedial actions. There are substantial differences between the ecological requirements of people living in different parts of the world. Excluding a few small countries, the American lifestyle is the most demanding of planetary resources. Despite having only 5 percent of the world's population, Americans use 20 percent of the world's energy, eat 15 percent of the world's meat, and create 40 percent of the garbage. However, the United States is not the only country that indulges in overconsumption of planetary resources. The ecological footprint of most developed countries is greater than the ecological services available within their geographical boundaries because they pass on the burden of their lifestyle to other parts of the world. The consequence of this profligate lifestyle is degradation of air, water, and land—in the immediate neighborhood and also in distant lands. Pollution is the largest single cause of disease and death in the world these days, causing an estimated 9 million premature deaths each year.[143] In addition, pollution decreases the productivity of farmland, which endangers the lives of millions of people by decreasing agricultural output.

Air Pollution

Air pollution is the concentration of harmful gases or minute particles in the atmosphere that causes numerous health and environmental problems. The major gaseous pollutants include carbon monoxide, sulfur dioxide, oxides of nitrogen, ammonia, and ground-level ozone. Power plants, oil refineries, and industrial facilities that produce consumer goods are stationary sources of air pollution, while motorized vehicles, such as cars, trucks, buses, planes, and trains, are mobile sources of pollutants. Some air pollution also emanates from natural sources such as forest fires, volcanic activities, and dust storms, but their contribution is much smaller. The major air pollutants, their sources, and effects are given in Table 1.

The two largest sources of air pollution in the United States are the burning of fossil fuels used for power generation and the transportation sector. Power plants that use coal as fuel produce copious amount of carbon dioxide, and also sulfur dioxide, nitrogen oxides, particulate matter, and mercury. Motorized vehicles used for transportation are the single largest source of air

Table 1 Major Air Pollutants and Their Effects.

Pollutant	Sources	Effects
Carbon Monoxide	Incomplete combustion of fossil fuels from heating oil, gasoline, wood stoves	Poisonous gas interfering with the function of blood
Nitrogen Oxides	Motor vehicles; Commercial, industrial, and residential use of fossil fuels	Lung damage; Respiratory illness; Precursor of acid rain; Formation of ozone
Particulate Matter (PM2.5 and PM10)	Combustion of fossil fuels	Lung damage; Cancer; Productive of haze
Sulfur Dioxide	Burning of coal, especially high sulfur coal; Paper manufacturing; Metal smleting	Lung damage; Production of acid rain that damages trees, lakes, and soil
Volatile Organic Compounds (VOCs)	Fuel combustion, especially cars; Solvents; Paints	Formation of ozone that damages vegetation
Ground-Level Ozone	Reaction of nitrogen oxides with VOCs: NOx and VOCs emitted by cars, factories, and power plants reacting chemically in the presence of heat and sunlight	Breathing problems; Reduced lung functioning; Damage to crops and vegetation; Production of smog
Lead	Leaded gasoline; Paint; Smelters	Brain and nervous system damage
Mercury	Combustion of coal and fossil fuels	Liver, kidney, and brain damage
Formaldehyde	Internal combustion engines; Particle board, plywood; fiberboard; household products	Cancer

pollution and produce almost a quarter of all hydrocarbons in the air and more than half of the carbon monoxide and oxides of nitrogen. Sulfur dioxide is produced by vehicles that use diesel fuels. Cars and trucks also emit other hazardous pollutants, such as benzene, acetaldehyde, and 1,2-butadiene, which are responsible for almost half of the cancers caused by air pollution.[144] Traveling by airplanes is much more damaging to the environment because it releases substantial amount of the greenhouse gas carbon dioxide. Americans travel by airplane more than people in other countries, and one-third of the global air traffic takes place in the United States. Out of a total of 32.6 gigatons of carbon dioxide added to the atmosphere by all human activities each year, air travel contributes around one gigaton. Since takeoff and descent require significantly more energy than cruising at high altitudes, flying non-stop reduces the amount of released greenhouse gases. Supersonic flights, discontinued a few years ago but being considered by some airlines, will have a much greater effect on the environment because they burn five to seven times more fuel for each passenger.

The Clean Air Act was promulgated in the United States in 1963 with amendments in 1970, 1977, and 1990. As a result of this act, outdoor air quality has substantially improved due to the use of cleaner fuels and more efficient vehicles, and to limiting the use of coal in power plants and factories. These developments reduced the concentration of pollutants in the air by 17 to 44 percent between the 1990s and 2015.[145] However, a report released by the American Lung Association in 2014 found that 147.6 million people in the country—47 percent of the nation's population—live in regions where pollution levels makes the air harmful for human health.[146] According to the EPA, indoor air in houses can be more seriously polluted than the outdoor air in the developed world. Air pollution inside homes is created by emissions from carpets, plastic products, heating equipment, and paints. These pollutants are trapped in houses that are generally isolated from the outside atmosphere. Both indoor and outdoor air pollution increase the incidence of cardiovascular diseases, lung cancer, and the risk of acute respiratory infections.

On a worldwide basis, about 92 percent of people are exposed to air pollution that is considered unhealthy by the WHO. This is a major health hazard in countries that are rapidly expanding their industrial bases. Air pollution is the fourth highest risk factor of death globally, causing an estimated 4.2 million premature deaths each year, making it one of the world's leading causes of premature deaths.[147,148] More than half of these fatalities occur in China and India, the world's fastest growing economies. In major cities such as Beijing, the air pollution levels are 100 percent higher than the limits set by the WHO. In 2014, only eight of the seventy-four biggest cities in China passed the government's air quality standard.[149] In the northern region of China, it is estimated that air pollution has decreased life expectancy by 5.5 years and it causes 1.2 million premature deaths each year.[150] There are also nonfatal consequences of exposure to air pollution—a study involving 25,000 people in different regions of China suggests that long-term exposure to polluted air adversely affects the performance in math and word recognition tests of older people, which may be a precursor to dementia.[151] Recently, China's government started an Air Pollution Action Plan that prohibits an increase in the use of coal in three major regions along the coast. However, some scientists think that the plan does not go far enough to prevent further degradation of air in major industrial centers.

India, the second most populous country, also suffers from poor air quality and smog in many major cities. The pollution in India also causes the premature deaths of a large number of people. A study determined that air pollution claimed 1.24 million lives in 2017, about 670,000 from air pollution in the environment and 480,000 from household pollution.[152] The human toll will continue to increase in the coming decades as the country industrializes. The air pollution in some Indian cities is increasing at a faster rate than in China. New Delhi, India's capital, is enveloped in so much smog in winter months that incoming airplanes are directed to other cities. From 2002 to 2010, the air pollution level in Bengaluru increased by 34 percent—the highest increase in air pollution in the world thus far.[153] Likewise, according to a study by the WHO, fourteen Indian cities are among twenty of the most polluted cities in the world.[154] Besides India, many other cities in developing countries have poor air quality that causes sicknesses in a large number of people.

A report by the United Nations estimates that air pollution costs the world's advanced economies, including India and China, about US$3.5 trillion per year in lives lost and health issues. The monetary impact of air pollution to the advanced Organization for Economic Cooperation and Development (OECD) countries is estimated to be US$1.7 trillion.[155] Although cities in the developed world do not suffer from air pollution as much as those in the developing world, many of them still do not meet WHO standards and are affected by air pollution in subtle but dangerous ways. The American Lung Association estimates that more than 47 percent of Americans live in areas with unhealthy levels of ozone or particulate pollution, and tens of thousands of people succumb to early death due to exposure to pollutants in the air.[156] Ground-level ozone, or "bad" ozone, is not emitted into air but is created by chemical reactions between nitrogen oxides and volatile organic compounds (VOCs) in the presence of heat and sunlight. High ozone days have been reported in Los Angeles, New York City, Chicago, Las Vegas, and Philadelphia. Despite significant improvements, air pollution remains a pervasive public health threat in the United States.

Ozone and Black Carbon

Ozone plays a dual role in the atmosphere. In the stratosphere, located roughly 6 to 10 miles above the Earth's surface, the presence of ozone is beneficial because it protects life on Earth from the Sun's harmful ultraviolet rays. This good ozone is naturally produced from oxygen in the stratosphere, but it can be depleted by chemicals that are used in industrial processes, such as chlorofluorocarbons, carbon tetrachloride, and chloroform. These chemicals migrate to the stratosphere where the Sun's ultraviolet rays break them down to release chlorine and bromine molecules, which destroy the good ozone. Fortunately, international efforts aimed at restricting the use of these harmful chemicals have substantially reduced the amount of ozone-destroying chemicals in the stratosphere.

At lower altitudes, known as the troposphere, ozone is the main ingredient of urban smog and is an air pollutant. Ground-level ozone is produced when substances such as carbon dioxide, nitrogen oxides, carbon monoxide, and volatile organic compounds in air are exposed to sunlight, resulting in the formation of smog. These substances are produced primarily through the burning of fossil fuels in cars and factories. Some of these chemicals are also produced from the burning of wood or animal waste, often done in developing countries. Ozone in smog irritates the respiratory system and causes coughing, throat irritation, and an uncomfortable sensation in the chest. Exposure to this gas for six to seven hours each day inflames the lining of lungs. Damaged cells are shed and replaced, much like the way skin peels off after sunburn. While respiratory inflammation in normal healthy people causes coughing, nausea, and chest pain, repeated inflammation can result in significantly reduced lung function. Inhalation for extended periods can cause premature death in people with heart or lung disease. Children and people with asthma or respiratory diseases are especially vulnerable, even at low ozone concentrations. Animal studies suggest that ozone may also reduce the immune system's ability to fight off bacterial infections of the respiratory system.

Air pollution also decreases the productivity of farmlands. Ozone in air is particularly harmful for plants and vegetation because it injures them and reduces their output by causing leaf damage that stifles plant growth. More specifically, black carbon (soot)—produced by burning wood and incomplete combustion of fossil fuels—is deposited on the leaves of plants, thereby reducing the amount of sunlight reaching the vegetation. Emission of these pollutants is increasing rapidly in developing economies and is particularly egregious in China and India. A team of scientists from the University of California, San Diego, evaluated the effect of ground-level ozone on the productivity of farmlands in India, the second-largest producer of staple crops. They showed that the output of wheat and rice farms is significantly reduced by ground-level ozone. On the national level, about 3.8 million tons of wheat and 2.3 million tons of rice are lost due to air pollution in India each year, which is enough to feed 94 million people in the country.[157] A statistical model suggests that, averaged over the country, yields were up to 36 percent lower than they would have been without the emission of these pollutants. Some densely populated regions experienced a decrease of 50 percent in the yield of wheat from farms.[158] On a worldwide basis, the reduction in the output of soybeans, corn, and wheat crops may be equivalent to the loss of US$17–35 billion per year.[159]

Particulate Pollution

Particulate pollution, which consists of small particles that float in air, is particularly dangerous for human health. These particles are solid substances or liquid droplets that often contain a dangerous mixture of acids, metals, and toxins. Some particles are emitted directly from the source—diesel trucks, power plants, wood stoves, etc.—while others are formed by chemical reactions in the atmosphere. These particles are classified according to size: particles with diameters of 2.5 micrometers or less are known as PM2.5 particles, while those with greater diameters of up to 10 micrometers are known as PM10 particles. Smaller PM2.5 particles, one-hundredth the thickness of human hair, pose a greater threat to health. They can reach the deepest alveoli in lungs and decrease their capacity to perform the crucial biological function of purifying blood in the body. These particles are dangerous for humans at any concentration and can cause cancer, heart attacks, and strokes. Several scientific studies have linked these fine particles to premature death, aggravated asthma, chronic bronchitis, and decreased lung function.[160] Long-term exposure to PM2.5 may lead to plaque deposit in the arteries, causing vascular inflammation and hardening of arteries that may eventually lead to heart attack and stroke. While all PM2.5 particles pose a threat to human health, those made of pure carbon are more injurious than the rest. PM10 particles are usually dispersed within a few miles from the source, but the smaller and more dangerous PM2.5 may be carried by wind thousands of miles from the place where they are generated.

Vast quantities of coarse and fine particles can produce haze that reduces out-door visibility. In addition to the effect of these particles on human health, they also cause environmental damage by making lakes and streams acidic and decreasing the output of farmlands.

Air pollution is a serious problem in South Asian countries and China. Considering overall air quality, Bangladesh has the most polluted air, closely followed by Pakistan and then India. Of the thirty most polluted cities in the world, twenty-one were in India in 2019.[161] In that year, 1.67 million deaths were attributed to air pollution, or 17.8 percent of all deaths in the country. The deaths and disabilities due to air pollution cost India a GDP loss of 1.36 percent, valued at $36.8 billion. The main sources of air pollution in the country are industrial activities (51 percent), exhaust from vehicles (27 percent), and the burning of crops to clear the land for the next plantation (17 percent). In 2015, the Government of India launched the National Clean Air Program with the target of reducing the concentration of PM2.5 and PM10 particles in the air by 2024.

The People's Republic of China (PRC) is the world's leading emitter of greenhouse gases and mercury. An estimated 1.24 million deaths in China were attributable to air pollution in 2017, including 851,660 from ambient PM2.5 pollution, 271,089 from household air pollution, and 778,187 from ambient ozone pollution.[162] Air pollution is an important public health concern in China, with high levels of exposure to both ambient and household air pollution. Although the government is making a serious effort to decrease the sources of pollution, air pollution remains an important risk factor.

Sulfur Oxide and Nitrogen Oxides

Oxides of nitrogen are highly reactive gases containing nitrogen and oxygen in varying ratios. They are formed during the combustion of gasoline at high temperatures and are emitted from the exhaust of motor vehicles and from stationary sources such as electrical utilities and industrial boilers. They are also precursors for ozone and PM2.5 particles. Ground-level ozone is formed when nitrogen oxides react in the atmosphere with other pollutants. Nitrogen dioxide can irritate the lungs and lower a person's resistance to respiratory infections.

The main source of sulfur dioxide is the combustion of fuels that contain sulfur—usually coal and diesel oil. The health effects of prolonged exposure to this gas are similar to those of the inhalation of ozone and nitrogen dioxide that damage the pulmonary system and cause respiratory illnesses. Sulfur dioxide can also combine with nitrogen oxides and moisture in the atmosphere to produce acid rain, which leads to a range of environmental problems. In addition to being toxic for plants, acid rain dissolves and washes away calcium and other minerals from the soil, thereby robbing it of nutrients that are essential for the growth of vegetation. Increased acidity of bodies of water also endangers the survival of some forms of marine life. Acid rain also damages or washes away features on buildings and other materials, including limestone sculptures and paint.

Carbon Monoxide

Carbon monoxide is a poisonous gas that is formed by the incomplete combustion of fuels such as gasoline, heating oil, natural gas, wood, and charcoal. This gas is dangerous for health and is deadly at high concentration. In the United States, the exhaust from motor vehicles produces roughly 60 percent of the carbon monoxide produced in the country. Other sources include industrial boilers, waste incinerators, and wildfires. The concentration of carbon monoxide is higher in winter months because vehicles work harder and burn fuel less efficiently in cold weather. On winter nights, a strong inversion layer develops in the atmosphere, trapping pollution near the ground and preventing it from mixing with cleaner air above.

Air pollution does not remain localized in the regions where it is produced. A study by scientists at NASA showed that air pollution can flow through the planet's entire atmosphere in just a few months. Asia's high level of pollution affects the health of people everywhere and changes weather patterns throughout the world. Some pollutants can rise to six miles in the upper atmosphere and create or disturb cloud formation and storms in distant parts of the world.[163] For example, acidic lakes in Scandinavia have been linked to pollution from factories in the United States.

Water Pollution

Water pollution refers to the contamination of bodies of water by harmful pathogens or waste materials. The most common sources of water pollution are chemicals used in agriculture, industrial effluents, household chemicals, and waste from animal facilities. Water is contaminated by the imprudent discharge of chemicals into rivers, streams, and ground. Since water is a good solvent, many pollutants that are injurious to human health get dissolved in it. Once contaminated, purification of large bodies of water is very difficult. Agriculture is a major source of water pollution because conventional agriculture requires the application of insecticides and herbicides at regular intervals to eliminate pests and prevent the growth of weeds. Most of these chemicals are harmful for human health and have been linked to damages in the respiratory system, birth defects, and cancer. Rainfall and irrigation wash away these chemicals from farmlands and pollute the groundwater. Animal factory farms use numerous chemicals, including pesticides, antibiotics, and growth hormones, that also end up polluting local bodies of water. A toxic mix of the bodily fluids of animals with these chemicals is stored in large ponds, known as lagoons. Groundwater becomes contaminated with chemicals and pathogens that leak from the lagoons, and also when their contents are applied to farms as fertilizer.[164]

The number of chemicals used by industries that produce consumer goods is very large—some of them are used to facilitate the fabrication of goods, while others are used to improve their appeal to consumers. Textile factories use around 8,000 synthetic chemicals to convert raw materials into final products. Features of finished products that a consumer finds alluring, such as vibrant colors and sheen, are usually obtained by using chemicals that are often toxic for humans. Common industrial pollutants include polychlorinated biphenyls, alkaline agents, dyes, benzene, chlorobenzene, carbon tetrachloride, toluene, and many other toxic or poisonous chemicals. Effluents from factories contain a mix of these chemicals which contribute to the pollution of local bodies of water. There have been many incidents in the U.S. in which a large quantity of industrial waste was released into groundwater by factories.[165] Manufacturing facilities are not concerned with the toxicity of these substances, since their only emphasis is on improving the visual appeal of products to increase sales.

The transportation and usage of fossil fuels also pollute the groundwater in many ways. Small leaks from millions of households and filling stations have a large cumulative effect on the environment and groundwater. Oceans are often polluted by oil spills from tankers and fuel lines. The 2010 leakage from the Deepwater Horizon in the Gulf of Mexico spilled 4.9 million barrels of crude oil into the ocean. The harmful environmental effects of the spill persist to this day.[166]

A typical American home contains a large number of chemicals that are harmful and toxic to humans or interfere with the productivity of the environment. These products include cleaners, paints, solvents, insecticides, stain removers, deodorizers, and cosmetics. Air fresheners and deodorizers usually contain the carcinogenic formaldehyde, while window cleaners contain diethylene glycol—a harmful chemical that causes damage to nervous, urinary, and digestive systems. A large portion of chemicals applied to home lawns and gardens, such as pesticides, weed killers, and fertilizers, eventually ends up polluting the groundwater. Many discarded drugs and household chemicals sent to landfills in household trash also end up polluting the groundwater.

Water pollution and lack of access to clean water is a major problem in many developing countries, including the two most populous countries—China and India. Seventy percent of the freshwater resources in India are contaminated and it ranks among the lowest of all countries in terms of water quality. Unplanned growth has led to the use of water bodies as dumping grounds for household sewage and industrial effluents. Water shortage and pollution are also serious problems in China—much worse in the northern parts of the country than in the south. Clean water is not available to a large proportion of China's population; an estimated 40 percent of the country's rivers are seriously polluted.

Land Degradation

Land degradation refers to the loss of productivity of land or soil due to human activities. It includes a wide variety of land conditions such as nutrient depletion, desertification, salinization (buildup of salts in the soil), erosion, compaction, and encroachment by invasive species. The Global Assessment of Soil Degradation commissioned by the United Nations Environment Program estimates that nearly 2 billion hectors—22.5 percent of the agricultural land, pasture, forest, and woodland—have been degraded since the mid-20th century.[167] While the primary cause of land degradation is pressure from the increasing population, it is also exacerbated by drought, climate change, poor water management, and farming practices that require immediate returns at the cost of long-term sustainability. Since the population of the world is increasing at a rate of about 80 million per year, loss of the productive land will decrease its capacity to provide food for the human population.

A broad classification of lands may be as follows: farmlands that are used to grow crops, rangelands that are less productive and used to raise livestock,

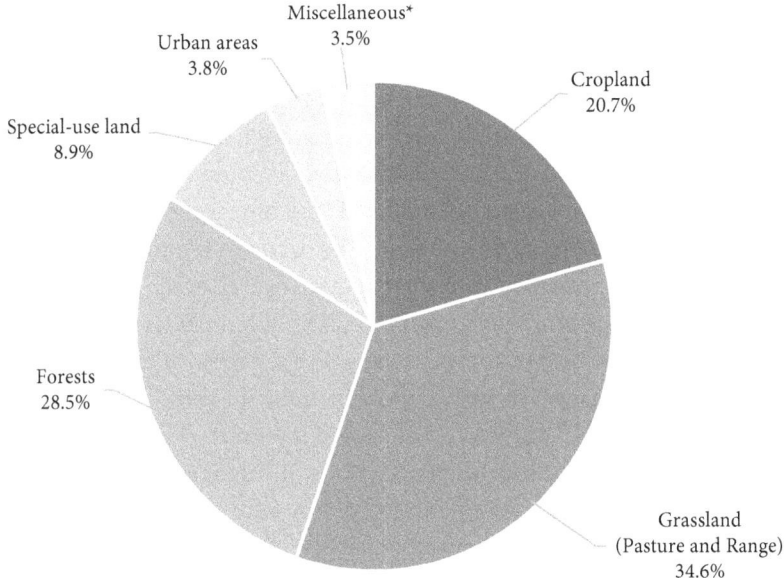

Figure 6 Land Use in the United States (excluding Alaska and Hawaii).
Note: *The category marked miscellaneous includes marshes, open swamps, rocky areas, deserts, and rural residential areas.
Source: USDA.

forests, deserts, unproductive lands, and areas used for human settlements underlying buildings and other structures. Satellite images show that a total of 3.4 million acres (1.38 million hectares), about 11.2 percent of the land, is used to grow crops and an additional 8.6 million acres (3.5 million hectares) is used as pasture for livestock.[168] Distribution of land use in the United States is shown in Figure 6.

Degradation of Farmlands

Most of the food crops grown in the U.S. are produced using industrial farming techniques. In this method, which is practiced in most developed countries and is becoming increasingly popular in the developing world, a single crop is grown over a large area. Corn, wheat, soybeans, cotton, and rice are all grown in this way in the U.S. Most of the work on farms, such as tilling the land, application of chemicals, and harvesting, is done with machines. Industrial farming depends on the regular input of water and application of fertilizers and other chemicals at prescribed intervals. In this method of farming, the land is tilled before planting and the crop is irrigated at regular intervals. Tilling the land brings nutrients and topsoil to the surface from where they are washed away by irrigation. Fertile soil contains about 40 tons of organic matter per acre (100 tons per hectare) in the form of decaying leaves, stems, and roots of plant products.[169] It is also inhabited by a diverse population of earthworms, insects, arthropods, and microorganisms. These organisms break down dead plants and animal tissues to form humus, the dark and crumbly carbon-based portion of soil that greatly enhances its productivity. Topsoil is very important for agriculture and is considered the foundation of human civilization. Industrial farming causes a loss of topsoil because the organisms that make it fertile die off and the nutrients are washed away by frequent turning of the soil and irrigation. Generating 3 centimeters of topsoil is estimated to take 1,000 years. If current rates of degradation continue, the world's topsoil could be gone in about sixty years.[170] Degradation of farmlands results in a loss of productivity and reduction in the output of farms.

To compensate for the loss of nutrients, and because modern varieties of grains have a greater requirement of nutrients, farming requires frequent application of synthetic fertilizers. Applying such chemicals for long periods leads to the accumulation of salts in the top layers of the soil. Frequent irrigation brings these salts to the surface, where they remain as water evaporates. Excessive salinity in the roots of plants makes them less vigorous and lowers the yield of crops. In extreme cases, it may create a salt crust on the surface of soil, making it impossible to grow anything at all. Excessive salinity is becoming a serious problem in many highly productive agricultural areas of the world. According to estimates by the United Nations, the world loses 2,000 hectares of soil each day, an area larger than Manhattan, due to excessive buildup of salts in farmlands.[171] Salt-degraded regions have been formed in some of the most productive farmlands of the world, including the San Joaquin Valley in California, Murray-Darling Basin in Australia, Yellow River Basin in China, Indo-Gangetic Plains in India, Indus Basin in Pakistan, and Aral Sea Basin in Central Asia. More than 40 percent of China's arable land has been degraded, according to the official news agency, Xinhua.[172] The global economic loss due to this phenomenon is estimated to be $27.3 billion per year.[173]

Industrial farming depends on the application of numerous chemicals to the crops at regular intervals, including fertilizers, pesticides, and herbicides. These chemicals are petrochemical products that contribute to global warming. The runoff of fertilizer that is not absorbed by plants into coastal seas causes algal blooms, eventually creating dead zones devoid of all forms of marine life. Most pesticides and herbicides are also injurious to human health. While the use of synthetic fertilizers adds the chemicals that are essential for the growth of plants (nitrogen, phosphorus, and potassium), most agricultural products need additional chemicals, such as magnesium, calcium, sulfur, and zinc, to develop proper nutritional value. Hence, the produce of modern farms is often less beneficial for health than earlier varieties that contained these chemical compounds.

Degradation of Semiarid Regions

A large part of Earth's surface, roughly 40 percent, is classified as semiarid. These lands have limited rainfall and their soil has low organic contents. They are not suitable for growing crops on an intensive basis but still have some capacity to grow vegetation. The land area in the world classified as semiarid is roughly twice the area that is used to grow crops. The native vegetation of these regions consists of shrubs and perennial grasses that stabilize the soil. Healthy rangelands provide many ecological services and benefits, including forage, carbon sequestration, biodiversity preservation, and storage for water runoff from higher elevation. Semiarid regions are inhabited by about 2 billion people in the world. The largest area stretches from the Sahara in Africa to parts of China. In addition, most of Australia is classified as semiarid, as well as much of the western regions of the United States. Semiarid lands are fragile ecosystems that are extremely vulnerable to degradation because their freshwater resources are limited, and they have shallow topsoil with low biomass. Inappropriately using these lands removes the vegetative cover and permanently degrades the soil. Global warming may further increase the degradation of these regions.

Even though the shortage of water and nutrients in soil makes it difficult to grow crops on a large scale in semiarid regions, some edible items can still be grown on a limited basis. The primary use of land in these regions is to raise livestock, hence they are also known as rangelands. While these lands can support some livestock, the increasing demand for meat forces farmers and ranchers to stock more animals than the land can support in a sustainable manner. As the size of the livestock herd increases, the perennial grasses decline because they are eaten by the farm animals. The native vegetation in these lands produces fewer seeds and does not quickly recover from damages caused by livestock. Continuous overgrazing removes all useful plants, leaving the land sparsely covered with unpalatable weeds. Removing vegetation also exposes the ground to intense solar radiation, leading to the loss of residual water by evaporation. The fertility of soil rapidly decreases when the vegetative cover is removed. When the land is compacted by the hooves of cattle, the soil cannot easily soak up the rainwater. Severe soil compaction, erosion, and decreased fertility now affect many cattle-ranching areas, including those in the American West, Central and South America, Asia, Australia, and sub-Saharan Africa. The United Nations Convention to Combat Desertification estimates that one-third of the semiarid land in these regions is moderately or severely degraded.[174]

Continuous degradation of land in semiarid regions may lead to desertification, a process that is characterized by erosion, loss of groundwater, and disappearance of native vegetation. This process converts productive semiarid lands into unproductive deserts. There is a fine line between arid land and desert, once the land is severely degraded it may be difficult to bring it back to productive state. Desertification is a slow, ongoing process that now threatens 10 million square miles (26 million square kilometers) of land in the world—just over 60 percent of the area of rangelands. The UN Food and Agriculture Organization estimates that it is affecting 168 countries and is one of the greatest environmental challenges of our time. About 43,000 square miles (70,000 square kilometers) of arid lands become deserts each year, an area about the size of Ireland. The arid and semiarid lands of the Western and Southwestern United States are highly vulnerable to desertification. The primary cause of desertification is the removal of vegetation. Human activities that denude the land of vegetation include overgrazing by farm animals and cutting down trees for firewood or industrial purposes. Variations in rainfall patterns caused by climate change also contribute to desertification by interfering with sustained growth of vegetation. Wildfires—increasing in both frequency and range in many parts of the world due to global warming—also denude the land of vegetation.

In earlier times, dry lands were mostly inhabited by nomads who moved with their herds of animals. Their constant movement prevented the land from becoming completely denuded. The use of boreholes and windmills these days allows livestock to stay in the same area throughout the year. These locations were formerly grazed only when the seasonal pans held water. Farm animals kept in semiarid lands all through the year eat the last remaining plants, leaving the soil unprotected from natural elements like sun and wind. Providing adequate drinking water to the growing number of farm animals has contributed to the massive advance of deserts in recent years. Once desertification has started, it triggers a chain reaction. Water in the soil evaporates in the absence of a vegetative cover, salt levels in the ground increase, and the ground hardens. Desertification also destroys the local ecosystem and causes the extinction of vulnerable species.

Advancing Deserts

One billion hectares of dry lands in Africa and about 1.4 billion hectares in Asia are affected by ongoing desertification. But it is not just a problem of developing countries. North America has the highest proportion of arid lands that may succumb to desertification. Five countries of the European Union are also facing desertification; some of the most vulnerable countries are in the former Soviet Union.[175] The Sahara Desert is expanding into bordering countries, creating problems for millions of people. In Central Asia, scientists estimate that 8,000 to 10,000 square kilometers of desert is created every year where cattle once grazed, or lakes existed. The degradation of semiarid lands causes an economic loss of US$6.3–10.6 trillion per year.[176]

Advancing deserts have taken a huge toll in China, which has always been arid. Currently, about 25 percent of China's landmass is classified as desert and the situation is getting worse—old deserts are advancing and new ones are forming. Sand dunes from the advancing desert are less than 100 kilometers from the capital Beijing. It is estimated that 40 percent of China is in danger of becoming wasteland. This development will make millions of people ecological refugees, thus creating enormous social and economic challenges. The dust from China's sandstorms often covers South Korea and Japan and even reaches the west coast of North America. Dust storms from the Gobi Desert in Asia and the African Sahara are responsible for respiratory problems in distant places, including North America.

Desertification is also a serious problem in India. According to the Indian Space Research Organization, about 69 percent of the land in the country is dry, making it vulnerable to water and wind erosion, salinization, and waterlogging. Land degradation causes loss of productivity and slowly converts it into wasteland. An estimated 105 hectares of land, about 32 percent of the total land in the country, is losing its productivity and slowly turning into deserts. Land in India is precious because it supports 17 percent of the world's population on only 2 percent of the global territory. It is estimated that land degradation is whittling away India's gross domestic product by 2.5 percent every year.

While 20 percent of the Western United States could be termed true, natural desert, another 20 percent has been so thoroughly and incessantly grazed by livestock that it has taken the appearance of a desert. In its Global 2000 report, the Council on Environmental Quality noted that improvident grazing has been the most potent desertification force in terms of total acreage—352,000 square miles or 910,000 square kilometers—in the United States. Livestock grazing is the most widespread land-management practice in Western North America. Seventy percent land in that part of the country is grazed, including wilderness areas, national forests, and even national parks. The U.S. Soil Conservation Service estimates that 410 million acres of public lands—21 percent of the land area of the country outside Alaska—are in unsatisfactory condition.

Climate change accelerates the rate at which deserts are growing, but desertification also contributes to climate change because the carbon dioxide sequestered in soil in the form of organic matter is released into the atmosphere when living soil is converted into a desert. In recent years, precipitation has declined in the Mediterranean region, southern Africa, and parts of South Asia. Southwestern states of the U.S. also faced a severe shortage of rainfall during the last few years. Heavy rainfall after a period of drought further reduces the fecundity of the land. During periods of drought, the surface layers of the land become fragmented. A flash flood then washes away nutrients from the soil. A short period of heavy rain does not recharge the ground with water that was lost during periods of drought.

A study by the Woods Hole Research Center concluded that the Amazon rainforest is at an imminent risk of being turned into desert.[177] If the 90 billion tons of carbon stored in that forest were to be released into atmosphere, it would have disastrous consequences for the world's climate, increasing global warming by an estimated 50 percent.[178] Extreme changes in one part of the planet affect even distant lands. Dust from the Sahara Desert drifts far away to the Americas, impacting weather patterns on this continent. Similarly, the water cycle in the Amazonian rainforests has an impact on weather patterns in Europe. Desertification directly affects the lives of about 2 million people who depend on those lands for livelihood, but it indirectly affects people living in all parts of the world.

Ubiquitous Plastics

Any discussion of land or water degradation is incomplete without a discussion of the role of plastics—the most ubiquitous substance in our daily lives. In one form or another, these petrochemical products are everywhere around us: in our homes, cars, clothing, furnishings, and food containers. The production of plastics has continuously increased over the last fifty years. A total of about 8.3 billion metric tons of plastics have been produced so far. An additional 300 million metric tons are produced each year, which is roughly the weight of the entire human population. The production of plastics is still increasing at the rate of about 9 percent per year, as manufacturers and consumers find new uses for them. About half the plastics produced are used in disposable applications—products that are discarded soon after use. The amount of plastic manufactured in the first ten years of this century is more than the total amount produced in the 20th century. Since these products do not degrade easily, billions of tons of plastics will keep accumulating around the world. Plastic debris can now be found everywhere, from the depth of oceans to the peaks of the highest mountains.

There are reasons for the enormous popularity of these products. Plastics are inexpensive, lightweight, strong, durable, corrosion resistant, and have excellent chemical and thermal properties. They can also be manufactured to have many different physical properties and find numerous uses because of their versatility. At least twenty types of plastics with widely different characteristics are made these days. The low cost and flexibility of plastics has resulted in the development of single-use, throwaway products that have a very short usable life, after which they are added to the already humongous amount of trash. Since the degradation of plastics in nature may take a few centuries, plastic made anywhere in the world, from the time when the first plastic was synthesized in 1907, is still present somewhere, polluting land or waters.

It has been estimated that in the United States more than 1 billion plastic bags are used every year, and 1,500 water bottles are discarded every second.[179] According to the Container Recycling Institute, 100.7 billion plastic beverage bottles were sold in the U.S. in 2014, about 315 bottles per person. One million plastic bottles are bought every minute in the world. Single-use plastic water bottles and bags are a major source of pollution. When deposited in landfills, plastics remain there for centuries, often leaching harmful chemicals in the groundwater. An estimated 17 million barrels of oil are required each year to produce water bottles, enough energy to fuel more than a million vehicles in the U.S. for one year. It takes more than twice the volume of water to produce a water bottle than the water contained in it.[180] Of the enormous amount of plastic produced, only about 9 percent of the discarded amount is recycled.[181]

About 8 million tons of plastic enter the oceans each year. It has formed the Great Pacific Garbage Patch, also known as the Pacific Trash Vortex, which extends from the west coast of North America to Japan. While some trash floats on the water, heavier debris sinks meters beneath the surface. Most of the debris is made of broken bits of plastic bags, bottle caps, plastic water bottles, and Styrofoam cups. About 80 percent of these objects come from coastal regions of North America and Asia, and 20 percent originates from boaters, offshore oil rigs, and stuff dumped by cargo ships.

Ocean plastics pose a threat to a wide variety of marine animals. Floating plastic pieces kill millions of sea creatures every year because they get entangled in nets or ingest fragments of plastic or plastic dust, causing liver or kidney abnormalities.[182] The aquaculture industry uses large amounts of plastic in floats, nets, lines, and tubes. The National Oceanic and Atmospheric Administration estimate that plastic in the ocean kills 1 million sea birds, 100,000 marine animals, and a very large number of fish each year. At the current rate, the weight of plastic in oceans will be more than that of fish by the year 2025.[183] Packaging accounts for the use of 40 percent of plastics, more than used in any other activity.

Some chemicals in plastics are absorbed by human bodies and cause hormone disruptions. Health officials have expressed concerns about the exposure of infants and children to these chemicals. Two of the most common chemicals are bisphenol-A (BPA), found in polycarbonate bottles and the lining of food and beverage cans, and di-2-ethylhexyl phthalate (DEHP) found in vinyl flooring, wall coverings, and food packaging. These chemicals are absorbed by the body and disrupt hormones and the endocrine system. Studies have found a correlation between exposure to these chemicals and harmful effects on human health, including early sexual maturation, decreased male fertility, and aggressive behavior. Exposure to these chemicals has a cumulative effect that increases with the passage of time.[184] Some plastics also contain DDT and PCB—two extremely toxic chemicals.

Recycling is often suggested as a superior method of getting rid of waste. However, recycled plastic is of inferior quality because sorting systems are not perfect and the quality decreases due to the mixing of different types of plastics in the process. When plastics are recycled, they are melted down, emitting harmful fumes. For this reason, most of the plastic to be recycled is shipped to developing countries like China and India, where environmental regulations are lax. Since the quality of recycled plastic is inferior to that of the original product, it is often turned into sturdier products like plastic lumber, patio furniture, roadside curbs, benches, and truck cargo liners. The fraction of plastic that is recycled is very low—only about 8.4 percent of the post-consumer plastic was recycled in the U.S. in 2017. The proportion is somewhat greater in Europe, where 26 percent of the used plastic was recycled.[185]

Policies of United States Governments

Richard Nixon was the first president to enact laws with the ambitious goal of improving the environment. The laws passed during his administration include the Clean Air Act (1970), the Clean Water Act (1972), and the Endangered Species Act (1973). He also established the Environmental Protection Agency to restrict air pollution throughout the country and to clean up hundreds of streams and rivers. These laws helped to improve the environment in the country and were somewhat improved by succeeding presidents. President Ronald Reagan tried to whittle some of these laws down but was mostly unsuccessful in those attempts because of opposition from the democratic Congress. President Obama enacted numerous laws to help the environment, including the control of greenhouse gases from power plants, factories, and new vehicles, and issued new rules to regulate fracking on public lands. Since ozone is harmful for human health, its limit was lowered from 75 ppb to 70 ppb.

During the term of the Trump administration, drastic changes were made in the U.S.'s environmental policy and many environmental rules were rolled back. Under his administration, the EPA promoted the interests of industries that they were supposed to regulate instead of protecting human health and the environment from their excesses. The list of environmental rules that were rolled back is very large and includes changes in emissions from the fossil fuel industry, the Clean Water Act, and the ban on dangerous pesticides and other chemicals. Trump's administration eliminated rules and regulations that the industries found cumbersome, regardless of their effect on human health and the environment, and dismantled most of the major climate and environmental policies enacted by the previous administration by calling them "unnecessary and burdensome." His administration officially reversed, revoked, or rolled back nearly 70 environmental rules and regulations.

The Obama-era Clean Power Plan was replaced with the Affordable Clean Power Rule that gives the authority to states to create their own rules for coal-fired plants. Obama's signature plan established a rule for reducing nationwide carbon emissions by 32 percent below 2005 levels. Under the law passed by Trump's administration, states have the option to impose looser restrictions that allow utilities to emit more greenhouse gases and other pollutants. According to various analyses, this change will cause an estimated 36,000 deaths and as many as 630,000 additional cases of respiratory illnesses each decade.[186] In January 2018, the repeal of the decades-old air emission policy opposed by the fossil fuel companies was announced. The EPA said it was withdrawing the "once-in always-in" policy under the Clean Air Act, which dictated how major sources of hazardous pollutants are regulated. Under Trump's administration, major pollution sources such as coal-fired power plants were to be reclassified as "area sources." The Trump administration withdrew clean air policies opposed by fossil fuel companies and weakened the Clean Air Act's enforcement of many pollutants such as arsenic, lead, mercury, and other toxins that are injurious for human health. Funding for programs that addressed the issue of air pollution was reduced.[187] In a stark departure from Trump, President Biden has promised that his administration will work to achieve a 100-percent clean energy economy with the goal of reaching net-zero emissions no later than 2050.

The Way Forward

The environment is a precious resource that is essential for our well-being and survival. Increasing the demand for ecosystem services without proper consideration of sustainability has brought us to a precarious state. A continuation of this trend will have drastic consequences for the whole world. The public's awareness of these issues is important so that individuals can take actions within their power, and force authorities to take appropriate corrective steps to prevent the progression of dangerous events.

The challenges faced by people, both individually and collectively, include cleaning up air and water and rejuvenating the productivity of oceans and farmlands. Air pollution already causes the premature deaths of millions of people, and contamination of groundwater causes sicknesses and deaths of a large number of people. As individuals, the first step we can take is to reduce or eliminate the use of chemicals that are harmful for human health or the environment and replace them with things that do not degrade the ecosystem and are not injurious to humans. The long list of such chemicals includes detergents, paints, stain removers, solvents, pesticides, and herbicides used in lawns and gardens. Individuals also contribute to the downward spiral of environmental degradation by driving gas-guzzling vehicles and dumping chemicals in trash and sewage. In general, we must develop a greater sensitivity for the environment. Even though the Earth is very large, it is not unlimited, and most degradations of the environment have a cumulative effect.

The production, storage, and consumption of food has multiple effects on the environment. Since industrial farming depletes the topsoil—a precious resource of farmlands that is essential to produce crops—it is important to switch to organic farming as much as possible. Consumers can assist in this process with their demands. Since a substantial portion of the produce of farms is lost in the digestive system of farm animals, decreasing the consumption of animal-based foods, or giving them up altogether for a vegan lifestyle, will reduce the pressure on farmlands and rangelands, thereby improving the sustainability of the food-production systems. Likewise, decreasing the wastage of foods, estimated to be about 30 percent, will further reduce the requirement for food and decrease the pressure of intensive farming.

While most consumers make decisions based on the price of items, it would be useful to consider their environmental impact as well. In the ultimate analysis, the environment is degraded by the demands and requirements of consumers, often propelled by industries. The increasing population also causes a greater demand for consumer products. Environmental degradation and pollution affect people in all parts of the world, either directly by harmful pollutants or by the loss of ecosystem services. Pollution is not generally confined to one region and slowly permeates to distant lands. Although their effects are more pronounced near the source, the environment of the entire world is slowly degraded because of the cumulative nature and migration of most pollutants.

There are many things that are outside the control of individuals and need governmental actions. Unless regulations exist and are strictly enforced, factories will dump waste materials into the local environment. Every stage of the use of coal—from mining to eventual combustion—pollutes air, land, and local bodies of water with harmful chemicals. Large commercial animal farms (CAFOs) produce enormous amounts of waste that often contaminate the soil and local bodies of water. Many factories still use fossil fuels for their operation. In the absence of strict enforcement, they will continue to contaminate the environment. Awareness of these issues by the public may force the administration to strictly enforce regulations on polluting industries.

Much of the pollution in developing countries is created by industries that manufacture items for the developed world. The demand for items at the lowest price and the competitive nature of the market forces manufacturers to find the place where the manufacturing costs are the lowest. In addition to cheap labor, this also requires that the manufacturing country absorbs environmental degradation without any compensation. Since multinational corporations do not have a commitment to stay in one country, the search for the cheapest place of production generates a competition between developing countries to entice these production facilities. In the ultimate analysis, wealthy consumers are partly responsible for the environmental degradation in developing countries. In addition, people living in developed countries are not completely protected from pollution generated in distant lands because massive pollution in one area will slowly degrade the environment everywhere. Consumers in the developed world can reduce environmental degradation by purchasing products that have not degraded the local environment.

FOUR

· ·

DEPLETION OF PLANETARY RESOURCES

There are numerous planetary resources that benefit us in many ways. Some of them are essential for our well-being while others are needed to maintain our lifestyle. Their exploitation greatly increased during the last few decades due to technological developments and the pressure of increasing population. We have been using the assets of the planet without considering its effect on sustainability, with the result that the biosphere is being degraded at a rapid rate. This trend is exacerbated by technological developments and results in a downward spiral of the depletion of nature's endowments required for our well-being.

The continuous exploitation of planetary assets is done with the implicit belief that the Earth is very large and will continue to provide for our needs forever. With this mindset, forests are razed for immediate benefits, oceans are used for dumping waste materials, and marine lives are decimated almost to the point of extinction. The immense diversity of the planetary flora and fauna that developed in the ecosystem over millennia, an important heritage of mankind, is being greatly reduced in favor of a few selected varieties that are useful at present.

Although the Earth is very large, the stock of resources that we need is finite, and their continuous depletion will lead to their exhaustion with serious consequences for humanity. However, this fact runs contrary to the profit motive of major corporations whose existence depends on a continuous increase in demand from the public, which is achieved with the help of advertisements and inducements of various kinds. Almost by definition, the time scale for businesses is short and they have to show profit without considering the long-term implications of their actions. For this reason, the culture of consumption, which runs counter to the long-term survival and welfare of humanity, has been assiduously developed by businesses.

Deforestation

Forests play an important role in preserving and maintaining the ecosystem and provide an essential service to all living creatures. Trees and plants absorb carbon dioxide through the process of photosynthesis, use carbon for their growth, and release oxygen that is necessary for the survival of almost all forms of life on Earth. This process also provides the added benefit of decreasing the concentration of carbon dioxide in the atmosphere. In addition, forests provide many ecosystem services such as purification and storage of water, enrichment of soil, moderation of climate, and support for an astounding variety of plant and animal life. They cover more than 30 percent of Earth's land and provide food, medicine, and fuel to more than a billion people. About 80 percent of the world's land-based species of plants and animals live in forests. Forests also preserve biodiversity and allow it to evolve in response to natural changes in the environment by providing habitat to species that are resistant to this change. In addition to purifying water and air, they play a critical role in mitigating climate change by soaking up carbon dioxide that would otherwise contribute to climate change. Forests provide habitat and livelihood to millions of people in all parts of the world—about 50 million people in the world have jobs that depend, directly or indirectly, on forests.[188]

Unfortunately, forests are being cleared at a rate of 18 million acres (7.3 million hectares) per year for some immediate gains, undermining the long-term benefits of this vital resource. Most forests are cleared to create room for cattle ranching, to accommodate urban sprawl, or to produce specialty items such as palm oil. Deforestation is occurring at rapid rate in the key emerging economies of Nigeria, Indonesia, and Brazil. Forests contain about 200 tons of carbon per acre, which is immediately released when they are burned. In addition, an equivalent amount of carbon is usually stored in the soil in the form of decayed biological materials and roots, which is slowly released into the atmosphere when the forest cover is removed. These facts make deforestation a major contributor to climate change. According to the Global Forest Resources Assessment prepared by the FAO, deforestation releases nearly a billion tons of carbon into the atmosphere each year.[189] Up to a fifth of global greenhouse gas emissions originate from deforestation and associated events.

The largest driver of deforestation is industrial agriculture in which large areas of natural forests are burned or cleared to raise cattle, to grow feed for farm animals, or to grow cash crops such as palm oil or soy. When deforestation is carried out by burning the forest, it releases, in addition to carbon dioxide, carbon monoxide and black carbon from the incomplete combustion of plant materials. These two substances are highly injurious to human health and may be carried to distant lands by winds. Forests are also cut down for industrial logging, to create building materials, and to make consumer products like office paper, tissues, books, magazines, and packaging materials. Mining for metals and building roads through forests also results in the destruction of forests, as does coal mining, which often requires the removal of forests from the top of hills that contain this ore. Deforestation sacrifices the long-term benefit to the ecosystem of forests for short-term gains.

Although all forests provide numerous benefits to humans and the ecosystem, rainforests are much more productive than others because the dense vegetation and frequent rainfall in these regions provides habitat for an astonishing variety of plant and animal life and provides many other benefits to mankind. Tropical rainforests are particularly useful for the global ecosystem because almost half of all terrestrial species of plants, animals, and insects live in these regions. The Amazon rainforest is the largest such forest in the world, covering about 8 percent of the landmass of the Earth (5.5 million square kilometers) in Brazil, Paraguay, and Argentina. The area covered by this forest is more than half (53 percent) of the area covered by all rainforests in the world. The rainforest in Africa covers 27 percent of the total, and those in Asia and Oceania cover the rest, about 20 percent.

Unfortunately, the Amazon rainforest is being destroyed at an increasingly rapid rate—almost 10,000 square kilometers of the forest was razed in 2019.[190] Besides plants and animals, the Amazon rainforest is inhabited by about 450,000 indigenous people who will become refugees if the forest is razed. President Jair Bolsonaro of Brazil has dismantled the environmental regulations that protected the forest, and the destruction of the rainforest has greatly accelerated under his leadership. Environmental advocates blame the policies of Bolsonaro for emboldening illegal loggers, ranchers, and land speculator to clear the forest. Clearing the rainforest of South America will have repercussions throughout the world. The land that is cleared by removing the forest is mainly used for cattle ranching and the cultivation of soybeans. The forests in Southeast Asia, Malaysia, and Indonesia, are cut down to grow palm trees because palm oil is in great demand as the most widely consumed vegetable oil, used in a great variety of consumer products, in addition to its use as a cooking oil. Chinese demand for wood is causing a loss of forests in many countries around the world. The rapidly growing Chinese furniture industry is supported by razing forests in Congo and Cameroon, Central Africa, and Indonesia.

Loss of Marine Life

The oceans cover about 71 percent of the Earth's area. Throughout history, humans have been awed by such a large expanse of water. It provides us with numerous benefits, including food, transportation, and recreation. Oceans also play a role in regulating the climate and weather patterns of many regions. For example, the North Atlantic current, or Gulf Stream, along with similar warm air currents, keep Ireland and the western coast of Great Britain a few degrees warmer than the eastern regions. The vastness of oceans made people believe—until recent times—that they can treat them as an infinite sink for waste products and sweep it for fish and other marine animals without risking a collapse of useful species.

The per capita consumption of fish in the world has been increasing steadily, from an average of 9.9 kg in the 1960s to 11.5 kg in the 1970s, 12.5 kg in the 1980s, 14.4 kg in 1990, and finally reaching 17 kg in the year 2011.[191] The yearly consumption of fish and shellfish in the United States is 21 kg (47 lbs.), which is considerably greater than the global average. The amount of fish consumed in different countries varies widely. For example, the average consumption of fish in African countries is very small compared to the rest of the world. Fish consumption has increased rapidly in China, from 13.9 kg per capita a few decades ago to 26.1 kg in recent years. In general, the consumption of seafoods is greater in region close to the shores.

There are several reasons why the consumption of fish is increasing more rapidly than other types of food. Fish is a high protein, low fat food that is also low in carbohydrates. Fish and fishery products represent a valuable source of protein and essential micronutrients. Oily fish, such as salmon, trout, sardines, tuna, and mackerel, also contain the beneficial omega-3 oils: EPA and DHA. Plant-based sources of omega-3 oils, such as flaxseeds, walnuts, canola oil, and avocado, have somewhat shorter acyl chains than fish oils. Although the human body has enzymes to convert plant-based omega-3 oils to EPA and DHA, the efficiency of these processes is low, and hence oily fish is generally considered to be a better source of these beneficial substances. Fish have more protein and less fat and carbohydrates because these creatures do not have to keep their bodies at a fixed temperature—they acquire the temperature of the surrounding water. Also, since they are naturally buoyant, they do not have to spend energy in locomotion and do not need muscles to support themselves. Fish meat is white—not red—because it does not contain myoglobin, the iron containing protein in muscles that imparts a reddish color to meats.

Declining Stock of Wild Fish

Before the introduction of large fishing vessels in the 1950s and 1960s, fishing was limited to coastal regions and the ocean was able to replenish the harvest of marine life. The new large vessels, however, travel thousands of miles and catch fish almost anywhere in the world. More recently, even larger ships with sophisticated radars, sonar tracking devices, and global positioning systems allow every part of the ocean to be explored by modern fishing fleets. Yet, the total weight of wild fish caught has been declining since 1994 at the rate of 700,000 tons per year because overfishing has decreased the net productivity of oceans.[192] Using a meta-analytical approach, Daniel Pauly and colleagues at the University of British Columbia estimated that the large predatory biomass of fish and seafood today is only about 10 percent of the preindustrial levels.[193] A study by Christopher Costello of the University of California, Santa Barbara, also presented a dismal picture of the state of the world's fisheries. The vast majority of useful fish populations have been depleted well below sustainable levels in many parts of the world, and the rate of decline is accelerating.[194] Another study by the FAO found that 85 percent of the global fish stocks are overexploited or depleted. Large areas of seabed in the Mediterranean and North Sea now resemble a desert, with no significant marine life.

To satisfy the increasing demand for seafood, the industry has adopted practices that are accelerating the collapse of edible fish by destroying the habitats of marine organisms and damaging the chain of life that is essential for the long-term survival of many species. Most commercial fishing these days is done with fishing trawlers. In this method, a net is pulled through the water behind one or more boats. The net, which catches all marine lives in its path, is known as a trawl. The fish caught in the net that do not belong to the target species are known as bycatch and are usually discarded. The method of fishing that causes most damage to the marine life is bottom trawling, in which a large net with heavy weights is dragged across the sea floor, scooping up everything in its path. The bottom trawlers these days are more sophisticated and powerful than earlier versions. They use vast, 30-ton nets that have metal doors and chains to hold them down as they are dragged across the seabed. Bottom trawling not only indiscriminately catches every form of life that it encounters but also destroys the habitat of creatures that use the ocean floor for food and shelter. Life at the seafloor may take a very long time to recover from such devastation.

Pollution of Seas

Human activities such as dumping waste products and industrial effluents have adversely affected the productivity of seas. A large portion of pesticides and other chemicals applied to farmlands ultimately ends up in the oceans. Oil spills from huge supertankers and other facilities have become common events with substantial impact on marine life. It is estimated that the Deepwater Horizon oil spill of 2012 reduced the production of wild fish by 100,000 metric tons.[195] Many industries in coastal areas consider the ocean to be an inexhaustible sink, too large to be adversely affected by human actions. But many pollutants are so dangerous that they either kill aquatic animals or render them toxic and unfit for human consumption. Industrial chemicals that are somewhat less potent stay in the food chain of marine creatures and are eventually ingested by humans who eat seafood. For example, mercury and heavy metals are accumulated in high concentrations in both fish and mollusks. Some species of fish are found to contain significant levels of methylmercury, PCBs, dioxin, and other environmental contaminants; hence, fish and other seafood are a major source of exposure to these harmful substances. PCBs and methylmercury stay in the human body for long periods and damage the organs of those who consume substantial amounts of contaminated fish.

Of all toxic substances in fish, mercury has received the greatest attention because it is present at various levels in all types of fish and other aquatic organisms. The biggest sources of mercury are coal-fired power plants. When coal is combusted, any mercury present in the ore is evaporated in the atmosphere. The amount of mercury present in coal is very small on a proportional basis but the amount of coal used in power plants is so large that a power plant releases a significant amount of mercury into the atmosphere from its smokestacks. Although it is possible to capture the mercury with scrubbers before it is released into the atmosphere, most companies are reluctant to incur the extra expense. The mercury in the atmosphere eventually ends up in the sea through rivers and streams. Once mercury enters a waterway, microscopic organisms in water convert it to a form called methylmercury. It is then ingested by marine creatures, increasing in concentration as it moves up the food chain. The species of fish that have the highest levels of mercury include tuna, sword-fish, sea bass, and bluefish. Humans are particularly vulnerable to the effects of this compound because it is a neurotoxin that interferes with the brain and the nervous system.

Dead Zones in Coastal Seas

The continental shelf is the portion of the sea that is closer to land and much shallower than the open oceans. It extends into the sea about 45 miles (73 kilometers) on average and its average depth is a few hundred meters. The physical and chemical characteristics of these regions are more variable than that of the open sea since they are influenced by seasonal variations in coastal lands. The combination of ample nutrients and shallow waters creates a favorable environment for aquatic life; hence a great variety of aquatic organisms thrive in the continental shelves. The surface area of these regions is only 10 percent of the areas of seas and oceans, yet they provide more than 90 percent of the world's harvest of fish and shellfish. Compared to the teeming life and high productivity near the continental shelves, the rest of the oceans are largely barren, mainly inhabited by large fish and a few species of mammals.

Industrial farming has a deleterious effect on marine life in coastal regions. Excess synthetic fertilizers that are not absorbed by the roots of plants are washed away by irrigation, with a substantial portion finding its way to the sea through rivers and streams. Fertilizers and biological waste trigger the bloom of some types of algae. As the algae rot and die, they remove dissolved oxygen from the water. Since the oxygen content of water is a crucial factor for sustaining all forms of marine life, bodies of water with excessive amount of nutrients and little oxygen become "dead zones" that cannot support higher organisms such as fish, clams, lobsters, and oysters.

Hypoxia (a deficiency of oxygen) is becoming more and more frequent worldwide, leading to the creation of the dead zones. In 2012, Robert Diaz, professor at Virginia Institute of Marine Science, identified 405 dead zones in coastal waters, covering an area of 95,000 square miles, about the size of New Zealand. The largest dead zone in the U.S., in the region where the Mississippi River ends in the sea, covers more than 8,500 square miles.[196] This number of dead zones is probably an undercount since large parts of Asia, Africa, and South America have not been properly studied. Diaz estimates that the actual number of dead zones in the world is more than 1,000.[197] A more recent study finds that the dead zones have quadrupled in size since 1950, while the number of very low oxygen sites in coastal waters has multiplied tenfold.[198]

Damages Caused by Aquaculture

While the number of fish caught in the wild has declined, the total worldwide production of fish has increased at a rate of 3.2 percent per year because the production of fish from aquaculture or farm-raised facilities has increased at an average rate of 8.8 percent during the three decades from 1980 to 2010, and is still increasing roughly at the same rate.[199] The global aquaculture production continues to grow, albeit more slowly than in the 1980s and 1990s. Aquaculture involves raising one species of fish, often genetically engineered, in enclosed bodies of water under controlled conditions. Salmon, carp, catfish, tilapia, and also invertebrates such as shrimp, clams, mussels, and oysters are raised in these fish farms. Far-eastern countries produce the bulk of fish raised in farms. China is the largest aquaculture producer, accounting for about 70 percent of the world's production. Eighty percent of the shrimp sold in the United States originate from aquaculture farms in Asia, mainly Thailand, Vietnam, and Cambodia. However, fish farms are now increasing in number in many countries of Europe and the Americas.

In a typical aquaculture facility, fish eggs that have been genetically modified are raised in a hatchery. The young fish are transferred either to inland ponds or to sections of the sea separated from the main body of water by nets. Farms in Asian countries often use inland ponds close to the sea to raise shrimp and tilapia, whereas they demarcate an area within the sea with nets to raise salmon. The success of fish farming depends on the availability of nutritionally suitable feed. Some of the species raised in aquafarms, such as tilapia, oysters, and carp, are omnivores that can also survive on a diet of grains, while others, including shrimp, salmon, and bass, are carnivores that require fish products in their diet. Fish farming also requires copious amounts of water. For fish raised in inland ponds, the water must be changed on a regular basis to prevent waste material from accumulating to toxic levels. Cages in coastal waters pollute the local environment to such an extent that they must be relocated every few cycles.

To increase the production and profit, the density of fish in these enclosures is kept very high—the maximum stocking density is limited only by excessive mortality in the population. Salmon farms usually raise as many as 50,000 individuals in each enclosure and are so crowded that a 2.5-foot fish spends its entire life in a space the size of a bathtub. Trout farms are even more crowded, with up to twenty-seven full-grown fish in a bathtub-size space. Such crowded conditions are conducive to the growth of microscopic pathogens that grow at very low levels, or not at all, in the wild. Bacteria and viruses growing in these enclosures sometimes reach epidemic proportions and wipe out the whole fish colony. Disease outbreaks have caused a loss of production in many aquaculture operations around the world. The pathogens invade the fishes' bodies through their gills and, unlike farm animals, signs of sickness in individual fish are not easily spotted by the farms' operators. The spread of disease becomes obvious only when many fish begin to die. In many respects, aquaculture is similar to raising livestock in industrial animal farms because the marine creatures pollute the local environment and require input from distant places. Even under optimum conditions, a significant part of the fish feed is not eaten and ends up on the ocean floor. The area around the aquaculture farms is polluted with tons of fish feces, antibiotic-laden fish feed, and fish carcasses. Accumulation of these substances sometimes causes the ocean floor around these facilities to rot, eliminating all useful forms of life from these regions.

Salmon farming pollutes the local environment and causes more problems per pound of fish than any other type of aquaculture. Millions of salmon escape from cages each year, disrupting wild salmon populations by introducing diseases and parasites, and competing with them for their habitat. Faster maturation and more aggressive farm salmon may initially deprive the smaller and more cautious wild fish of food and shelter, but ultimately the farm fish fail to survive because they are bred to live in controlled environments and do not have the hardiness of wild varieties. Mating between these two groups introduces vulnerabilities in the wild stock, further endangering their survival.

Many of the highly prized species raised in aquaculture farms, such as salmon, bass, trout, and shrimp, are carnivores that require fishmeal and fish oil as food and cannot survive on a herbivorous diet. These fish are fed the byproducts of trawler fishing that indiscriminately hauls all kinds of aquatic organisms into their nets. After separating and processing the desired species as food for humans, everything else is ground up and turned into fishmeal for aquaculture facilities. Fishmeal and fish oils are the only available sources of the highly unsaturated fatty acids that are both essential nutrients for all carnivorous fish and a key reason for their health appeal. It takes three to six pounds of wild fish to produce one pound of farmed fish. Around 11 million tons of wild fish—about 12 percent of the total haul from seas and rivers—are fed to farmed fish. Fishmeal used to fatten high-value salmon or bass often includes herring, sardines, and mackerel that could also be directly consumed by people.

Since all carnivorous fish and shrimp are net protein consumers rather than producers, aquaculture is hastening the depletion of wild stocks of life in open seas. In search of feed for farm-raised fish, some operators have begun harvesting krill, the tiny shrimp-like creature found in cold waters. These organisms feed on phytoplankton and are near the bottom of the food chain. They form an important component of the diet of larger fish and other marine animals. Large-scale harvesting of krill is bound to have an adverse effect on the survival of all forms of life in the seas. This process sacrifices sustainability in favor of short-term gains.

The level of beneficial omega-3 fatty acids varies significantly in farmed salmon, depending on the type of feed given to them. According to the International Fishmeal and Fish Oil Organization (IFFO), the level of omega-3s is declining in farmed salmon. In 2008, 3.5 ounces of farmed salmon contained 2 to 2.5 grams of beneficial oils, EPA and DHA, down from 3 grams about three years earlier. Since 2008, it has come down further to about 1.5 grams for each serving of farmed salmon.[200] To some extent, the oil present in all kinds of farmed fishes depends on the feed given to them. Catfish, tilapia, and shrimp are very low in fat and therefore low in omega-3 fatty acids, although they are good sources of protein. A vegetarian diet fed to fish that can survive on such feed produces lower amounts of omega-3 fatty acids in their flesh, thus decreasing their beneficial effect for human health. Several studies have also found surprisingly high levels of toxic chemicals such as PCBs and dioxin in farmed fish.[201]

The stock of wild fish is decreasing, and their habitats are being destroyed. It has been suggested that unless urgent measures are taken, we may be the last generation to obtain food from the oceans. About 85 percent of the global fish stocks are depleted, fully exploited, or overexploited.[202] As a result of bottom trawling, large areas of seabed in many parts of the world now resemble a desert in that they do not contain any form of marine life. Although aquaculture seems to be filling the gap in the harvest of wild fish created by their diminishing stock, there are ample reasons to believe that fish farming cannot be sustained for a long time. A future collapse of ocean fisheries, as has been projected by many studies, would have an adverse impact on the production from these facilities.

Aquaculture operations have other undesirable effects as well. Fish farming requires both land and water, two resources already in short supply in many parts of the world. Industrial production of shrimp and salmon is carried out mostly in the tropical regions of developing countries. In many of these regions, mangrove forests and wetlands have been destroyed to make way for fish farms. The fish produced are beyond the means of the local population and are produced to be exported to wealthier countries. By occupying precious territory in coastal regions, these farms often deprive the local community of fishing opportunities.

Decreasing Biodiversity

Biodiversity consists of an incredible variety of plants, animals, and marine life that developed over billions of years of evolution to meet the challenges of different environments and living conditions. It would be hard to enumerate all the complex ways in which the organisms living in various ecosystems interact with others of the same or different species to form a highly intricate web of life. However, the burgeoning human population and developments in technology have increased the reach and impact of humanity to the farthest corners of the world, frequently endangering the survival of many species. Many varieties of plants and animals have become extinct in the modern age, and the populations of species that have not become extinct have greatly diminished. This "sixth extinction" is caused by humans, hence this era is known as the Anthropocene (*anthropo* meaning *human*) Epoch.[203]

There are two kinds of diversities in the animal kingdom: species diversity and genetic diversity. While species diversity refers to different species, genetic diversity is the diversity of characteristics within the same species. A healthy ecosystem performs vital functions beneficial for humanity in many ways, including nutrient cycling, carbon sequestration, pest regulation, and pollination. The fecundity of the soil is maintained by the vegetative cover and numerous insects and microorganisms. Insects break down waste materials and convert them into useful forms. Earthworms contribute to the fertility of the soil by turning it, and spiders and some insects grind the organic matter and leave behind enriched droppings. Algae provide organic matter and incorporate atmospheric nitrogen into the soil. In a healthy biosystem, the waste produced by human activities is broken down and turned into useful forms by insects and other organisms. Natural vegetative cover in water catchments helps to maintain the hydrological cycle, regulate and stabilize water runoff, and acts as a buffer in extreme events such as floods or droughts. It is estimated that pollination by honeybees produces $1.6 billion worth of agricultural products per year in the United States, and other pollinators, such as different varieties of bees, insects, and birds, contribute up to $6.7 billion.[204]

Marine life provides direct benefits to humanity without the need of cultivation. Wild fish and other forms of marine creatures provide the main source of protein to about 1 billion people. Modern pharmaceutical science recognizes the importance of chemicals found in plants. It is estimated that about 25 percent of modern drugs contain a compound that was initially extracted from plants, such as aspirin, codeine, atropine, and vincristine, although these drugs may now be chemically synthesized. Animal venoms and toxins have been found to be useful in making drugs to treat diseases such as hypertension, chronic pain, and diabetes. Investigations of toxins from snakes, lizards, fish, snails, octopuses, and scorpions are being actively pursued in many laboratories. The Economics of Ecosystems and Diversity (TEEB), an organization backed by the UN and various European governments, estimates that the markets in many sectors, including the pharmaceutical, biotechnology, agricultural, and food and beverage industries, depend on natural genetic resources. It is impossible to put a price tag on the benefits that we acquire from biodiversity, but the cost of useful items runs into trillions of dollars.[205] In addition, biodiversity stores resources that are of no immediate use now but may become useful in the future.

The International Science Policy Platform on Biodiversity and Ecosystem Services (IPBES), an organization of the United Nations, issued a global assessment of the state of planetary ecosystem in May 2019.[206] It was the work of 145 experts from fifty countries that took three years to prepare and was the most comprehensive assessment of its kind. The report assessed changes in the planetary ecosystem during the past five decades and concluded that 1 million species of plants and animals are threatened with extinction. Nature is declining globally at rates unprecedented in human history. The IPBES chair, Sir Robert Watson, stated: "The overwhelming evidence of the IPBES Global Assessment, from a wide range of different fields of knowledge, presents an ominous picture. The health of the ecosystem on which we and all other species depend is deteriorating more rapidly than ever. We are eroding the very foundation of our economies, livelihood, food security, health and quality of life worldwide."[207] The biodiversity crisis spans the globe and threatens all ecosystems and species of animals.

The IPBES report finds that more than 40 percent of amphibian species, almost 33 percent of reef-forming corals, and more than a third of all marine mammals are in danger of extinction. Similar losses are predicted for insects and vertebrate species. More than 9 percent of all domesticated animals used for food and agriculture had become extinct by 2016 and at least 1,000 more breeds may disappear within a few decades. In general, the report finds that species of all kinds—mammals, birds, amphibians, insects, plants, marine life, and terrestrial life—are disappearing at a rate tens to hundreds of times higher than the average over the last 10 million years. The contributing factor to these changes is the fact that three-quarters of the land-based environment and about 66 percent of marine environment have been significantly altered by human actions. Before the IPBES report came out, the World Wide Fund for Nature (WWF) had estimated that there was an astonishing 60 percent decline in the size of the populations of mammals, birds, reptiles, and amphibians during the last forty years.[208]

Human activities have changed many ecosystems. Dead zones are spreading in the highly productive coastal seas in all continents. Many tropical forests are eerily silent because the insects have vanished. Semiarid lands are being converted into deserts in many parts of the world. There is an imprint of humanity almost everywhere—to the detriment of local flora and fauna. The area of the world that has been unaltered by humans is shrinking all the time. Climate change adversely affects the habitat of many species, and hunting and poaching further decreases the populations of numerous animal species. Pollution of all kinds makes life difficult for many animals. Marine lives are particularly affected by the plastic waste floating in the waters. The overall effect of the loss of plant and animal species may be almost as deleterious as that of climate change. Life on Earth is in peril. Sandra Diaz, one of the co-chairs of IPBES, says: "The evidence is crystal clear. Nature is in trouble—therefore we are in trouble."

The primary reason for the loss of biodiversity is the destruction of the habitat of plants or animals caused by encroaching human populations. In addition, the existence of some species is threatened by deliberate human action. The dodo bird was hunted into extinction and the number of wild elephants is precipitously declining because of the ivory in their tusks. The search for a quick solution without full environmental analysis also plays a significant role. For example, applying insecticides to prevent damage to crops by invasive species also kills beneficial insects such as honeybees. Pristine forests are razed to provide feed for farm animals or to be used as pasture, causing enormous harm to the creatures that lived in them. Similarly, expansion of suburban homes to regions not previously inhabited by humans often kills many forms of life.

Global warming also contributes to the loss of some species by altering their habitat. A substantial portion of Arctic and Antarctic ice has already melted, allowing the passage of ships in regions that were earlier covered with ice all year round. This has made life difficult for animals that reside in these regions, including polar bears, sea lions, and penguins. Higher temperatures also prevent the growth of algae that only grows in very cold waters and is at the bottom of the food chain for fish, birds, and marine mammals.

Some of the carbon dioxide in the atmosphere produced by the combustion of fossil fuels is dissolved in oceans, thereby increasing their acidity. Greater acidity, warming of oceans, and pollution are detrimental for the survival of coral reefs, which are dying in all parts of the world. The Great Barrier Reef, a natural wonder off the coast of Queensland in Australia, is in the grips of death and decay. Coral reefs are home to a very large variety of living creatures, including fish, turtles, eels, crabs, shrimp, sea urchins, and sponges. They have been compared to rainforests because of the wide variety of life that they support. When the reefs die, they will take with them about one-third of the marine biodiversity. Death of coral reefs due to pollution, increased temperature of waters, and greater acidity leads to the death of many of these creatures.

Applying chemicals to kill invasive or unwanted species of plant and animal life often has an undesirable effect on useful species. For example, insecticides belonging to the family of neonicotinoids are the most widely deployed insecticide these days. These insecticides, 10,000 times more potent than DDT, have been shown to persist in the soil for more than a decade, and are a major cause of the decline of bees and other beneficial insects. Typically, more than

90 percent of this pesticide stays in the soil for long periods and its concentration increases with each application. As such, it has been suggested that neonicotinoids may cause enormous damage to many forms of terrestrial life, including birds and animals that eat insects.[209]

Biodiversity may be crucial in these times of climate change because some subspecies may survive and even thrive under changed conditions. We are living in an age of homogenization of food sources. Out of the 30,000 species of edible plants in the world, humans now depend on fewer than a dozen for a full 80 percent of their caloric intake. Human diets have also become increasingly similar in all parts of the world. Staple crops like wheat, rice, corn, and soybeans have displaced regional crops, such as cassava and sorghum. A global diet that mostly consists of a handful of staple crops is highly vulnerable to disease, pests, or climate change that may destroy a substantial portion of the crop. An institutional commitment to biodiversity is important because it may be necessary to search for genetic traits that will survive under adverse conditions. In addition to food, biological systems also generate many useful materials. About a quarter of drugs in use these days were initially discovered from natural sources. Biodiversity in a healthy ecosystem may allow the evolution of new species in response to changing surroundings. In general, a healthy ecosystem is maintained by a continuous interaction between vegetative growth, insects, and animals. A loss of biodiversity deprives us of resources that are important for the sustenance of the ecosystem crucial for our survival in future.

To safeguard against the continuous degradation of the ecosystem, transformative changes are urgently needed. People and governments everywhere must realize that chasing economic growth without caring for nature will be detrimental for humanity. At this stage, people everywhere must realize that we have already done enormous harm to the ecosystem. While individuals should examine the effects of their actions on the environment and refrain from activities that harm other forms of life, countries should set aside more land for nature as protected areas. This protection does not mean that humans cannot go there but that the areas are protected from resource extraction and land conversion. Unless steps are taken to preserve biodiversity soon, the sources of food for humans will keep decreasing. Nearly one hundred organizations around the world have endorsed the goal of protecting half of the planetary resources by 2050, with an interim goal of protecting 30 percent by 2030 under what is called "Global Deal for Nature."[210]

Depletion of Mineral Resources

Technological developments in recent years have increased the need for minerals, some of which had very little use in earlier times. A few of these minerals, crucial for the performance of the electronic devices, are found in limited amounts and at the present rate of consumption their known reserves will be exhausted within a few decades. These minerals include indium, a soft, malleable metal, gallium, used in LEDs, and rhenium, used in the manufacture of jet engines. Rare earth metals, a group of seventeen chemical elements, are critical for the manufacture of many products, including computers, cell phones, magnets, fluorescent lights, and many advanced technology products. Rare earth materials neodymium and praseodymium are used to make powerful magnets that are essential components of many electronic devices. China has 37 percent of the world's reserve of rare earth metals, Brazil is second with 18 percent, and Russia is third with 15 percent of the total. The U.S. has only about 1 percent of the total known reserve in the world.

One compound that is crucial for our survival is phosphorus in the form of phosphates. Our bodies store and release energy by converting a crucial molecule to high phosphate (ATP) or low phosphate (ADP) states, a process that goes on all the time. The food that we eat produces ATP, which powers virtually every activity of cells and organisms. All physiological processes that require energy obtain it directly from stored ATP, thereby converting it to a low-energy state, ADP. It has been estimated that 10^{26} molecules undergo this conversion to store energy from food and release it whenever work is done, including minimal activities like breathing and pumping of blood by the heart. Agriculture throughout the world requires about 20 million metric tons of phosphate to maintain its productivity. There are very few natural deposits of phosphate in the world. It is primarily recycled—from human and animal waste to agricultural lands where plants absorb it. Regions of the U.S., Europe, and some other countries, where large amounts of fertilizers have been applied to the crops over the years, have excess phosphate in the soil which, unfortunately, runs into rivers and streams and eventually to the seas with irrigation and rainfall. Conserving phosphate in the food chain by composting and recycling is perhaps the only way to feed the increasing human population. Farm animals return only a small fraction of the phosphate that they consume in animal-based foods, yet a large amount of phosphate is required to grow feed for them. Hence, the consumption of animal-based food increases the requirement of this crucial mineral.

Effects of Globalization

Globalization involves free exchange of ideas, technologies, goods, and services in different countries around the world. One of its main objectives is to remove barriers to international trade. Developments in communication and transport technologies have facilitated and accelerated globalization during the last few decades. World trade plays an important role in the sharing of goods between nations and has grown five times in real terms since 1980. Its share of world GDP has risen from 36 percent to 55 percent over the same period.[211] The pace of globalization has somewhat slowed in recent years because of Brexit in the United Kingdom and Trump's "America First" approach in the United States. Now China is the strongest defender—and beneficiary—of globalization. Globalization was supposed to enable free trade between countries by reducing barriers such as tariffs, value-added taxes, and subsidies. However, the G20 countries have added more than 1,200 restrictive export and import measures since 2008 to advance their own agenda.

Globalization has—and continues to have—a substantial impact on the lives and livelihoods of a large segment of people in many regions of the world. The initial premise of globalization was that it will bring prosperity everywhere by breaking down barriers to trade. It has, however, created both winners and losers. A small section of the middle class in developing countries, such as China, India, Indonesia, and Brazil, has improved its financial status.[212] At the same time, the inflation-adjusted income of the top 1 percent of earners increased by more than 60 percent during the last two decades. Globalization has been good for the multinational corporations and Wall Street, but it has not been good for the middle class in America, which has seen jobs disappear or move to developing countries. The COVID-19 pandemic greatly increased the wealth of the richest people in the U.S. and the U.K. The profits during the pandemic were so immense that American billionaires could provide a $3,000 stimulus payment to every, man, woman, and child in the country and still be richer than they were before the pandemic began.[213]

Trade with China has a particularly large effect on the prospects of American workers. China joined the World Trade Organization (WTO) in 2001 and soon became the largest trading partner of the United States. President Clinton, while making a case for admitting China to the WTO, said that it would create a "win-win result for both countries." However, the U.S. trade deficit with that country increased from $84.1 billion in 2001, the year China joined the WTO, to $375 billion in 2017. The U.S. imported $506 billion worth of goods from

China in 2017 and earned only $130 billion through exports.[214] Trade deficit displaces the American workers who would have been gainfully employed with workers in other countries. Using an input-output model, the Economic Policy Institute (EPI) calculated that the trade deficit with China eliminated 3.2 million U.S. jobs between December 2001 and December 2013. Most of these jobs (75.7 percent) were in manufacturing industries.[215] These job losses were in every congressional district in the country, including the District of Columbia.

While globalization is moving industries away from the United States, workers in developing countries that have become manufacturing hubs are treated as slave labor. To maximize profits, multinational companies pay the lowest possible wages to workers, and subject them to coercive, abusive, unhealthy, and unsafe working conditions. Globalization has led to exploitation of labor. Prisoners and child workers are often used to work in inhumane conditions. Safety standards are often ignored to save on the cost of production. When workers in a country begin to organize and demand higher wages, the company moves its operations to other parts of the world. In the search for the cheapest place to manufacture goods, many companies are relocating from China to other countries.

Globalization has a deleterious effect on the environment in a number of ways. With its emphasis on greater consumption of goods, usually produced by factories that use fossil fuels either directly or indirectly, it increases the global burden of greenhouse gases. Corporations locate their factories at places where the environmental regulations are nonexistent or lax, with the result that these places suffer from massive pollutions from industrial effluents. With no commitment to sustainability, the local resources are exploited to the maximum extent. Air in these regions is also polluted from volatile organic chemicals used in the textile and fashion industries. The fear that major industries will move to other countries if they impose too many conditions forces most countries to lower their environmental standards to meet the demands of multinational corporations. After producing merchandise at the lowest possible cost, businesses sell their products in developed countries through advertisements and inducements of various kinds. Relatively low prices of consumer goods produced in developing countries and a continuous barrage from the media have given rise to a "throwaway" culture in the developed world that is bad for the ecosystem in numerous ways. It has also greatly increased the wealth of the superrich in the Western world.

The profits of multinational companies depend on finding or producing goods anywhere in the world that can be sold at a higher price, generally to customers in the developed world. With this objective, they search and appropriate items that may have an appeal in rich countries. They may also, through local subsidiaries, encourage the production of items that may be useful in Western countries, even though this process may deprive the local communities of essential items or despoil their environment. Shrimp and fish grown in aquaculture facilities for Western markets occupy mangroves and coastal areas in East Asian countries, thus depriving the local population of an opportunity to fish for food. When palm oil or coconut water becomes popular in the West, forests are cleared to produce these items without considering the effect on the environment. The popularity of quinoa as a wholesome food makes it so expensive in the local market that people who have depended on this grain as a major component of their diet for centuries cannot afford it. When customers in the U.S. realize that grass-fed beef may have nutritional advantages over grain-fed beef from feedlots, forests in Brazil are razed to produce this commodity.

Most flowers sold in the United States and the European markets are imported from Colombia, which has almost twelve hours of sunshine throughout the year. The United States Department of Agriculture checks flowers for insects, hence the growers douse the plants with pesticides and fungicides. The flower industry is a profligate user of water—it has been estimated that producing a single rose bloom requires as much as 3 gallons of water.[216] Primarily due to the demand of water by the flower industry, springs, streams, and wetlands are disappearing in Colombia. Workers in these fields are exposed to more than a hundred types of hazardous chemicals, many of which are toxic for humans and are also carcinogenic. Exposure to these chemicals by pregnant women workers often results in miscarriage, premature birth, and babies with congenital defects. Primarily due to the demand for water by the flower industry, groundwater levels have plunged in the country, and springs, streams, and wetlands are disappearing.

Globalization and technological developments have helped the rich to become superrich, greatly increasing the inequality in most countries, including the U.S. The salary of top executives increased by 997 percent between 1978 and 2014, while the compensation of a private, nonsupervisory worker increased by just 10.9 percent during those thirty-six years.[217] The financial industry executives and CEOs of companies, who belong to the 0.1 percent of the population, are the biggest beneficiaries and have seen their incomes reach astronomical numbers. The superrich and the corporations owned by them have hidden about $30 trillion—an incredibly large amount—in tax havens such as the Cayman Islands, Switzerland, Ireland, and the Netherlands. This practice allows the superrich to preserve their wealth while the host country of these corporations (including the U.S.) must educate and provide health care and other services to their workers. Social welfare schemes that help the indigent in the developed world are under great pressure because of national deficits, job losses, and other economic ramifications of globalization.

Large companies that have manufacturing facilities in developing countries have enough power to dictate the terms for keeping the factories in those countries and demand that environmental regulations be loosened, or resources made available to them without constraints. Corporations demand and ensure that developing countries keep producing primary products with little scope for economic development. The power of large multinational corporations allows them to exploit local workers and the environment solely for their benefit, which makes them resemble colonial powers of the previous era. Globalization forces poorer nations to surrender their powers and remain in the straitjacket provided to them. Multinational corporations (MNC) have become so powerful in many developing countries that they influence the political process and help pass laws, and even elect politicians, that support their interests. This may give rise to a new era of colonialism where these corporations control much of the world. The resources of some multinational companies already exceed the GDP of many countries. Partly due to globalization, inequality has worsened both internationally and within countries.[218] The rich businessmen in these countries act as conduits for the operation of MNCs and reap financial rewards. In all the so-called BRICS (an acronym for the names of the five major emerging economies: Brazil, Russia, India, China, and South Africa), the share of income going to the richest 1 percent of the population increased from about 5 percent to more than 20 percent during the last twenty years.[219] Globalization also leads to cultural homogenization in which the language, food, dress, and traditions of rich people slowly replace the cultures of all other lands. The traditions of indigenous peoples developed over millennia will succumb to the continuous onslaught from media. This loss is somewhat comparable to the loss of biodiversity and extinction of species.

The Way Forward

Consumers can help protect planetary resources by making judicious choices in the marketplace, and by electing representatives who consider saving the environment to be a priority. In addition to reducing the consumption of products that lead to the destruction of forests, consumers can make an impact by requiring companies to introduce zero-deforestation policies certified by a rigorous third-party certification program such as the Forest Stewardship Council. Consumers also have a role to play in protecting species that are in danger of extinction. For example, the endangered Chilean sea bass, bluefin tuna, and grouper should not be consumed. The Marine Stewardship Council (MSC) now labels seafood that is either caught or farmed in ways that takes the long-term stability of the harvested species into account, and also considers the livelihood of communities dependent on fisheries. Although some questions have been raised about their selection process, it appears to be a useful first step. In general, it is better to consume smaller fish that are lower in the food chain because they multiply rapidly.

The control of pests by homeowners and businesses often involves the use of chemicals that are harmful for human health and the environment. Consumers can play a major role in protecting their health and preventing environmental degradation by using only substances that are benign for the environment, even though they may not be as effective as poisonous chemicals and may require greater effort. Globalization is not beneficial for workers in rich countries or for workers in countries where the products are manufactured; it basically increases the wealth of the superrich. It is therefore useful to buy locally manufactured items and not succumb to the allure of products made in distant lands.

FIVE

· · · · · · · · · · · · · · · · · · · ·

OVERCONSUMPTION

Humans have always used ecosystem services for their survival and well-being. Technological progress has added many items to the basic needs of food, shelter, and necessities of life. Consumption becomes overconsumption when it is not sustainable—when the ecosystem cannot continue to provide resources for producing the goods and cannot absorb the waste generated by human activities. Overconsumption has far-reaching and cumulative effects on the entire ecosystem.

When factories began to mass produce many more items than people wanted, corporations made concerted efforts to increase the demand, as exemplified in a statement made by Paul Mazur, a Wall Street banker working for Lehman Brothers in the 1930s, "We must shift America from a needs to a desires culture. People must be trained to desire, to want new things even before the old had been entirely consumed. We must shape a new mentality in America. Man's desires must overshadow his needs."[220] This strategy was supported by successive governments. President Herbert Hoover told a group of advertisers and businessmen: "You have taken over the job of creating desire and have transformed people into constantly moving happiness machines. Machines which have become the key to economic progress."[221] The mass media was eminently successful in expanding the desires of the population to completely new territories as Americans happily complied and a new era of consumer culture was born.

The age of continuously expanding needs and desires has been in full swing for more than half a century and is now threatening the integrity of the ecosphere. There is a disconnect between our needs and desires. Humanity's ever-increasing demands lead to overexploitation of natural resources such as water, energy, fertile land, and minerals, and are causing serious degradation of the environment. Excessive use of fossil fuels is changing the climate, freshwater resources are dwindling, the stock of wild fish is decreasing precipitously, fertile lands are losing their productivity, and species are becoming extinct at a rate that is unprecedented in history. Sixty percent of nonhuman primates, our closest biological relatives, are now threatened with extinction.[222] Many experts think that we are currently on the brink of a sixth mass extinction that will lead to losses and declines of vertebrate populations.[223] Various plant and animal species that support our lives in ways that are not properly understood or appreciated face existential challenges. Even though the ecosystem has some resiliency, damages caused by humans during the last several decades have caused immense harm to the natural world from which it would be difficult to recover. Instead of giving a respite to the ecosystem, the damages are continuing at an ever-increasing rate.

In the third quarter of 2019 Americans spent $3.8 trillion on items for personal consumption.[224] On a proportional basis, the per capita personal expenditure is somewhat similar in other developed countries. Consumption is continuously increasing everywhere in the world. In developed countries, personal consumption is increasing at a rate of about 3 percent per year, while in developing countries, which started at a much lower level of consumption, the growth is about 5 to 7 percent per year. In addition, the worldwide population is increasing at a rate of about 80 million per year. These two factors—increasing consumption and growing population—continuously increase the burden of humanity on the ecosystem. In a search for resources, we scout the entire planet with tools provided by modern technology causing ecological damage by razing forests, despoiling pristine lands, and emptying oceans of marine life. Continuously increasing consumption, facilitated by technological developments, is likely to cause serious harm to the global ecosystem. The effluents from factories that produce an innumerable variety of goods frequently contaminate air, water, and land with harmful substances. The discarded products, after their intentionally designed short usable life, accumulate in the biosphere, interfering with the productivity and proper functioning of the ecosystem. We are slowly coming to the realization that the continuously increasing demand for resources cannot be met by the planet and consumption at this rate is not sustainable.

It is difficult to account for all planetary resources that are used by humans each year. However, activities of daily life frequently involve expenditure

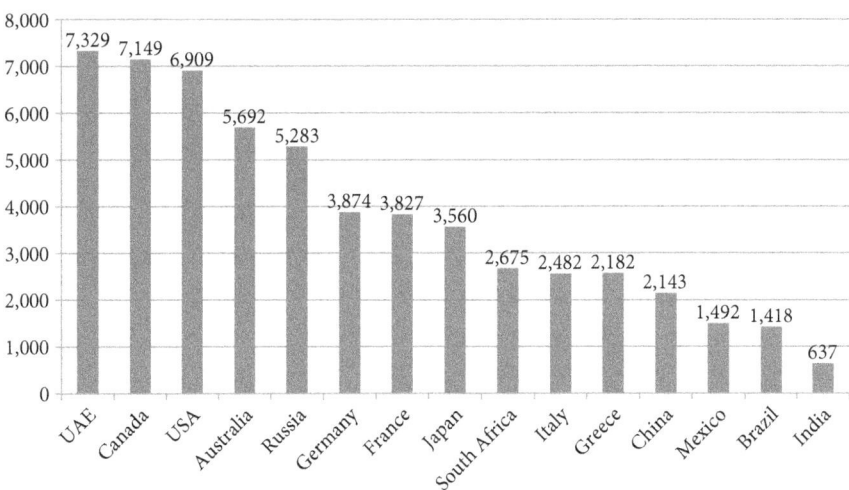

Figure 7 Per Capita Energy Use (kg of oil equivalent).
Source: World Bank.

of energy. A rough indication of the use of resources is indicated by the per capita energy consumption. Figure 7 shows the energy usage in a few selected countries. Since there are many sources of energy, such as hydroelectric, nuclear, wind, and solar, the total energy has been converted to the equivalent amount of oil. The per capita energy consumption of the U.S. and Canada is more than three times that of China, the most populous country.

Consumer Culture

Although consumption is an essential part of life, the emphasis on possession has given rise to a consumer culture that equates possession of items with the worth of a person and his status in society. It makes excessive consumption appear normal, even desirable. Consumption to maintain status or to satisfy vanity distorts priorities in life. While possessions are acquired to bring happiness, any feeling of joy is temporary because once an item has been acquired the next item becomes more alluring. The desire for continuously increasing acquisitions becomes an opiate that keeps people from worrying about important things such as the lack of a secure job, affordable health care, supporting worthy causes, saving for a rainy day, or paying off the student loans that are carried by 40 million individuals and amount to a staggering $1.48 trillion.[225]

By all metrics, Americans are the richest people in the world. An average American has more consumer products than people anywhere else. However, surveys have shown that consumption, in general, does not make people happier. A Harris poll found that only 33 percent of Americans said that they were happy.[226] A nationally representative sample of Americans found that only one component of consumption is positively related to happiness: leisure activities. All other expenditures, e.g., on personal care, food items, health care, vehicles, housing, and charity, are not significantly associated with happiness.[227] In fact, studies have shown that there is a connection between excessively materialistic outlook and increased levels of anxiety and depression.

In this age there is a complete disconnect between basic needs and desires; companies make an item and then create need for it with seductive advertising and other means. This way of thinking has become the norm in all developed countries and is rapidly making its way to the developing world. The newly emergent class of affluent people in China, India, and other developing countries tries its best to emulate the American style of consumption. A significant aspect of consumer culture is that the consumer is generally not interested in its antecedents—how it was manufactured or produced, the labor involved in these processes, the entire chain that brings it to the market, and the environmental effects from its production to eventual disposal. An implicit assumption is that the ecosystem has virtually limitless capacity to provide resources and absorb pollution and waste, hence a continuous expansion of production and consumption is considered to be an essential part of living.

Drivers of Consumption

Humans have an innate need for status and novelty. In earlier times, some people acquired a higher status in society simply because they were born into a privileged class. In 17th- and 18th-century England, use of tobacco, sugar, and tea declared that the user belonged to a privileged group. These days, the majority of people advertise their position in a social hierarchy with their possessions. Acquisition of valuable items—in both quality and quantity—announce one's position in the social hierarchy. Consumption is encouraged by the U.S. government because it constitutes 70 percent of the economy; the proportion is somewhat similar in other developed countries. One of the ways in which the government can encourage an industry is through tax breaks, as is done for home ownership these days. Advertisements for products in the mass media get a tax deduction, which encourages companies to spend large sums of money to promote their products. When the economy is in the doldrums, the government often tries to improve the situation by providing economic stimuli in the form of rebates and checks. Citizens are then exhorted to spend that amount to rejuvenate the economy. Government can also encourage consumption of an item by subsidizing it, thus lowering its price to the consumers.

The common dictum among industries is that one must grow or perish. Growth, in the common parlance of the industry, means increasing the demand of whatever the company produces. It could be said that growth by increasing consumption is an essential feature of the free-market capitalist system. The profit on capital requires economic growth—to be achieved by increasing demand for the product. This endless cycle ignores the ecological limits of the planet. In recent decades, the trend among consumers is toward bigger houses, more powerful and spacious cars, trendier fashions and accessories, and the latest electronic devices such as smart phones, three-dimensional televisions, internet-connected watches, and even devices that transpose the person into a world of virtual reality with mind-boggling experiences. The economist Thorstein Veblen coined the term "conspicuous consumption" in 1899 for consumption undertaken to make a statement to others about one's class, accomplishments, or success in life. This is a proper description of the state of things these days; possessions give the owner of these items a feeling of self-worth. Since barriers of nobility and aristocracy have broken down, wealth becomes the only metric for judging a person's status in society. While the really rich (1 percent) do not have to display their wealth with consumer goods, the status of middle-class persons is often defined by their possessions. This leads to a competition to possess trendy things before others. The urge to acquire things without thinking of the long-term effect on personal finances and the environment is extraordinarily strong for an average person. Online retailers have made it easier to buy things; many of them offer one-click shopping, which means that the whole process of buying takes minimum effort, freeing time to continue shopping for more items.

The Role of Advertising

Businesses have been phenomenally successful in converting people into consumers that are "constantly moving happiness machines" with the help of enticements and inducements of various kinds. Advertisement is the most important component for promoting the sale of goods and has become a major force in shaping the values and aspirations of the public. The amount of money spent on advertising reached a new high of $241 billion in the U.S. in 2019 and is expected to reach $250 billion within a few years.[228] This huge amount still does not include direct mail and phone marketing. With the increasing popularity of tablets and handheld devices, advertisers are spending greater amounts on these devices every year.[229] The amount of money spent by American companies on advertising is almost twice the amount that the government spends on education ($107.6 billion) and is more than the GDP of many countries. An average American sees 362 advertisements through radio, television, newspapers, and phone marketing per day. The number of messages and brand exposures to each person is more than 3,000 per day and may be much higher if it includes all signs that one sees in a grocery store, while driving a car, on computers or tablets, or on the labels of clothing and household items.

A considerable amount of scientific and consumer behavior research has been done to improve the effectiveness of advertisements. The main objective of direct advertisements is to attract the viewer's attention, which may require unusual behavior or even crazy antics. Advertisements appeal to people's emotions in different ways. Some of them, like those of the insurance industry, use fear to motivate people to buy their product. Many appeal to the desire for fun and pleasure, while others try to persuade people to buy their product by appealing to their vanity and ego. The fear of aging, or the appearance of looking old, has been deeply ingrained in our minds by advertisements. It is used to promote beauty products, and also some invasive procedures such as reconstructive surgery and Botox treatments to remove signs of aging from face or body. Women are made to believe that there are deficiencies in all parts of their bodies—hair, nails, eyelashes, etc.—that can be corrected by manicures, pedicures, surgery, and other procedures. The advertisements create feelings of anxiety and inferiority and suggest that submitting to corrective procedures or purchasing fashionable attire will fix everything. This approach was suggested by Charles Kettering of General Motors in 1920 when he said: "The key to economic prosperity is organized creation of disaffection."

In contrast to direct advertising, subtle advertising involves placing an object at a visible place in a television show or movie, particularly in association with a celebrity or well-known character. This form of advertising has been found to be highly effective. However, its opportunities are somewhat limited. It may even lose its appeal if its motive is obvious or if it is done very often. Another form of subtle advertising is association of the product with a celebrity. The shoes or apparel worn by a well-known actor or sports figure may persuade some people to acquire the same things. This technique is particularly useful in the fashion industry, where celebrities are paid huge sums of money to display or support some items. Publicity agents also use various psychological techniques to promote their products.

Advertisements aimed at children have been shown to be particularly pernicious because children lack the ability to resist commercial messages. When a child looks at a television advertisement, she assumes that all her peers are using the same product. While most dietary guidelines suggest that snacks, convenience foods, and sugary beverages should be consumed only occasionally and in limited amounts, these are the food items that are advertised most heavily to children. Food and drink companies spend $12 billion per year in advertising their products to children. Today's children, ages eight to eighteen, consume multiple types of media and spend more than forty-four hours per week in front of computers, televisions, and game screens. The American Psychological Association has concluded that advertisements of non-nutritious convenience foods are partly responsible for the increasing incidence of obesity in children.[230] Advertisements of various kinds make young people, particularly girls, very conscious of their image at a very young age. The simultaneous increase in obesity in almost all countries seems to be driven by changes in global food system from freshly prepared foods to highly processed items that are effectively marketed through advertisements and other means.[231]

Easy Accessibility of Credit

Extending credit is an essential requirement for maintaining consumer culture. The sale of goods would substantially decrease if people had to put down the full amount at the time of purchase. Businesses, including banks, are eager to extend credit to customers to facilitate the sale of goods. A typical customer only wants to know the monthly payment while deciding to purchase an article and not the extent to which the financing company is exploiting the situation. Easy accessibility of credit increases sales because it encourages impulsive buying. The average U.S. household credit card debt stands at $16,061, counting only those households that carry debt, and the total credit card debt carried by Americans is $931 billion.[232] If home mortgage and student loans are included, the total debt on the American consumers is about $11.86 trillion.[233]

Planned Obsolescence

Another method of increasing consumption is planned obsolescence—a business strategy in which objects are designed to become obsolete because they are no longer usable or trendy, and the consumer is forced to replace them and purchase the latest versions of the products. This is done in a number of ways. One of the methods is to make items that have short life spans or have expensive parts. One example is an ink cartridge with a very small capacity in a printer. A generation ago, washing machines and refrigerators were designed to last a lifetime. The technological features of washing machines made now have decreased their usable life to about a decade. The strategy of planned obsolescence is common in the computer industry. New versions of software can read files in the older versions but not the other way around. Thus, people holding on to older version can only communicate with those who have not upgraded to the latest version. On the hardware side, many programs become so huge with each new release that an older computer does not have the capacity to access them. Another method of increasing consumption is to print a very short expiry date on drugs or edible items, making consumers throw away perfectly usable products. The objective of corporations is to maximize profits by selling as many items as possible without any concern for the unnecessary loss of resources.

Perceived obsolescence—the feeling that one is using old equipment—is a powerful psychological tool that forces people to buy new equipment even when the old one is working properly. It is not just clothes that are in fashion one week and out of fashion the next week; the same applies to electronic gadgets, toys, and even home furnishings. A fair amount of overconsumption in the Western world is not done to fulfill basic needs but to make social statements or to fill some void in people's lives, such as anxiety or depression. According to psychologist Robert Putman, the consumer culture leads to the loss of friendship and neighborly support, and to the weakening of communities.[234] The happiness and satisfaction that people derive from possessions is transitory, resulting in even more unhappiness when the glamour of the new object wears out. At the same time, the obsessive relationship with material things jeopardizes human relationships.

Major Contributors

Although all industries try to increase the sale of their products, the approaches and methods that they employ to entice customers are somewhat different.

They contribute to the degradation of environment and exhaustion of resources in many ways.

Fossil Fuel Industry

The fossil fuel industry, which produces coal, oil, and natural gas, is the largest industry in almost all countries. The United States government gives out more than $20 billion to support the production of oil, coal, and gas each year. In 2015, President Obama had asked for a substantial decrease in these subsidies, but the final budget passed by Congress either reinstated the amount given to these industries or increased it because of intense lobbying by the agents of this industry. This preferential treatment is not limited to the United States; the total subsidy given to fossil fuel industries each year by the G20 countries is more than $400 billion.[235] The direct costs of the use of fossil fuels borne by the society are pollution, climate change, and land degradation, while indirect costs are adverse health effects such as asthma and cancer in the general population.

Fashion Industry

The fashion industry deals with all kinds of apparel—men's, women's, and children's clothing, and also household furnishings. The vast network of the fashion industry begins with the production of raw materials, such as cotton grown on agricultural farms or synthetic fibers produced from petrochemicals, and includes designing, manufacturing, distributing, marketing, retailing, advertising, and promoting the product until it entices customers to buy the final product. In monetary terms, it is the second biggest industry in the world, second only to the oil and power industry. The ultimate cost of all these processes is borne by consumers. The U.S. market is the biggest in the world. Americans spent about $400 billion on various types of apparel in 2015; the total for all countries was $2.4 trillion per year.[236] Fashion has been growing at a faster rate than the rest of the economy—during the last decade, the industry expanded at the rate of 5.5 percent annually.[237] Growth in this sector is driven by demands by individual consumers, hence advertisements and inducements of various kinds, including shock tactics, play an important role in promoting sales. Because of a continuous barrage from media, people consider it very important to follow the latest styles and feel awkward if they are not in step with the latest fashion. Those who acquire trendy things proudly display them, thus acting as agents of the industry.

An important driver of the fashion industry is the phenomenon of continuously changing styles—so-called "fast fashion." In earlier times, apparel retailers would change their displays three or four times a year, in keeping with changing seasons. These days, there are as many as ten or twelve changes in fashionable clothing each year. Some giant retailing stores have started upgrading their collection every week, or even twice a week.[238] Spring collections begin in the dead of winter with a markdown of winter clothing and undergo many changes by the time warmer weather arrives. Continuous changes in trendy clothing promote a relentless drive for overconsumption. The glamour and appeal of staying in step with the current fashion has been carefully cultivated in the minds of people by the industry. A continuous barrage of advertisements and pressure from peers has resulted in a large increase in the wardrobe of the American woman. In 1930, the average American woman owned thirty-six pieces of clothing, making nine complete outfits. Today, a woman has 120 items of clothing and adds an average of sixty-four pieces of new clothing to her wardrobe each year.[239] The world's resources cannot keep up with the increasing demand for throwaway fashions.

A significant feature of fast fashion is that the quality of apparel is not an important consideration since the only emphasis is on its appeal in the showcase. Manufacturers and retailers are well aware that the lifespan of these articles is very short. These items are often bought on impulse and frequently confined to the closet after a few uses, or without being used at all. The fashion industry maintains the allure of continuously changing trendy clothes by keeping the price reasonably low. The emphasis of manufacturers is on appearance and not on quality, which is achieved by using cheap materials. After the clothes have been washed once or twice, the dye will fade, the appearance will degrade, and the clothes will lose their fit and finish.

Most consumers do not realize that the apparel they buy degrades the environment and uses precious resources in all phases of operation—from their production to eventual disposal. Cotton, which is the basic fiber of roughly half of the clothes and textiles, uses more water than most other crops; it takes more than 2,000 gallons of water to produce 1 pound of cotton. Since this plant is highly susceptible to invasion by pests, growing cotton requires an application of copious amounts of pesticides, which eventually wash into the soil and pollute groundwater. Pests frequently develop resistance to pesticides, requiring application of newer and more powerful pesticides, further degrading the environment with even more harmful chemicals.

The most common and widely used alternative to cotton is polyester, which is a petrochemical product. The production of polyesters and other synthetic fibers is an energy-intensive process that uses crude oil and releases numerous pollutants and greenhouse gases into the atmosphere. These synthetic materials are like plastics and will stay in the ecosystems for centuries, harming many forms of terrestrial and marine life. Synthetic fibers also emit almost three times more carbon monoxide in their life cycle than cotton.

A total of 400 billion square meters of textile are produced annually. Since making articles of clothing from fabrics requires extensive cutting, an estimated 60 billion square meters are left on the cutting board. About 1.7 million tons of about 8,000 synthetic chemicals are used in dying and treatment of textiles to give them an appealing sheen and finish. Textile factories are responsible for about 20 percent of industrial water pollution with chemical effluents. Some of these chemicals are highly toxic, even carcinogenic. Fast fashion has a large pollution footprint; each step in the life cycle of clothing generates environmental hazards, making the fashion industry only second to the fossil fuel industry in its deleterious effect on the environment. Transportation of clothing over long distances—by land, sea, or air—adds to the burden of greenhouse gases in the atmosphere.

The eventual disposal of these items also adds to their harmful effects on the ecosystem. One way that people in developed countries get rid of their excess clothing is by donating it to charity. Since there is not enough demand for used clothing in developed countries, a large portion gets sent to developing nations. Although the donors feel good about it, many used clothes sent to Asian and African countries are not acceptable to people there because they are not designed to last long and do not conform to local styles. A report by Oxfam found that a large fraction of clothing reaching African countries was unsalable due to poor quality. Used clothing from developed countries also creates a relationship of dependency that discourages the growth of local industries because local producers cannot compete with the inexpensive clothing received from Western countries.

The other alternative to disposing of old clothing is to trash them. In less than twenty years, the volume of clothing Americans toss each year has doubled from 7 million tons to 14 million tons, which amounts to 80 pounds per person each year. The environmental effect of trashing textiles is equivalent to the carbon dioxide emission from 7.3 million cars.[240] The toxic and carcinogenic chemicals used to enhance the appearance of these articles leach from them in landfills and a significant portion ends up polluting the groundwater. Burning discarded clothing releases numerous toxins into the air. Trashing clothes is also expensive; nationwide, municipalities pay about $45 per ton for waste sent to a landfill. It costs New York City $20.6 million annually to ship discarded textiles to a landfill.

In addition to the environmental effects, multinational corporations meet the massive demand of inexpensive, trendy clothes in the Western world by hiring about 170 million child laborers in developing countries.[241] Only about 3 percent of the clothing is made in the U.S., the remaining 97 percent is made in other parts of the world—from Mexico to Vietnam and China. Child labor is unlawful in most countries, but the International Labour Organization has determined that the fashion industry is the biggest employer of child labor, where they are engaged in making furnishings and garments for consumers in Europe and the United States.

Industrial Livestock Facilities

The food that we choose to eat places a burden on the resources of the planet. An average American consumes 275 pounds of meat per year. In general, the consumption of meats and seafoods in developed countries is about three times that in developing economies, but the consumption in developing countries is increasing rapidly with increasing prosperity. Roughly 65 percent of grains grown in the United States are fed to livestock. Since the conversion efficiency of agricultural products to animal-based foods is very low, a high-meat diet puts a much greater burden on the ecosystem than a diet primarily based on agricultural products. The most important resource that is increasingly under stress in many countries is water. Animal-based foods require large amounts of water in all phases of operation, from growing feed and providing for the daily needs of animals during their lives to slaughtering facilities and shipments to consumers. Animal farms occupy a lot of land, either directly or indirectly, for growing feed for animals. The demand for animal-based foods often necessitates clearing rainforests in tropical countries.

Plastics

Plastics exemplify the throwaway culture that is causing enormous harm to the ecosystem. Plastics are now found almost everywhere because of their low price and excellent physical properties. It is almost impossible to find a place where there is no plastic in some form or the other. They are lighter in weight than metals or glass, the materials that they replace, and can be easily molded into any shape required by the application. There are at least twenty types of plastics with widely different properties. These polymers are inexpensive because they can be fabricated from fossil fuels by simple processes. The versatility and low cost of plastics allows manufacturers and consumers to use articles made of these materials with abandon. Since plastics do not undergo biodegradation for centuries, the litter of carelessly discarded materials creates an eyesore and a public health hazard. When deposited in landfills, they often leach harmful chemicals into the groundwater.

Globalization and Overconsumption

An important objective of globalization is to make goods from one part of the world available to people all over the globe. In principle, it could be beneficial to all parties if items are grown or produced where local conditions are more favorable and exported to other regions of the world. For example, rice can be more easily grown where rainfall or sources of water are plentiful and exported to relatively arid regions. In practice, however, a significant part of globalization involves manufacturing goods in developing countries and providing them to consumers in the developed world.

Globalization increases consumption, mainly by people living in developed countries. At present, 80 percent of the world's resources are used by 17 percent of the world's population belonging to the developed world.[242] With an emphasis on increasing consumption, sustainability is not an important consideration. In fact, manufacturing facilities set up by MNCs in developing countries use precious resources and also degrade the environment. When the local environment has been substantially degraded or the resources have been exhausted, MNCs simply move their operations to another country; thus the process continues and the environment is degraded in many regions of the world. The global ecosystem is a shared resource and degraded water, air, and land slowly affect the entire world.

International trade redistributes resources, not necessarily in an equitable way, with no concern for sustainability. It allows countries with high purchasing power to use the resources of poorer countries. World trade accelerates depletion of resources by making them available to distant lands. Resources of poor, low-consuming countries that will last for a long time, if used locally, are transferred to high-consuming countries. MNCs play an important role in facilitating international trade. They control, either directly or indirectly, the production of articles in developing countries and their distribution in affluent countries. Manufacturing facilities are set up in developing countries where labor is cheap and environmental regulations are either lax or can be easily circumvented.

In some cases, consumption of selected items in rich countries deprives people in less developed countries of basic necessities. Flowers, fruit, and shrimp are often produced in developing countries for export to richer countries. This is done at the expense of produce that may be consumed locally. Globalization forces some developing countries to produce articles that are desirable to Western consumers instead of growing produce for local consumption. As an example, farmers in the Ica region of Peru grow the thirsty crop of asparagus to supply to the Western markets throughout the year. Laborers in asparagus fields cannot even afford to feed their children, while the exporters and foreign supermarkets benefit from the sale of this luxury vegetable.[243] About 1.5 billion roses and other cut flowers sold in the United States are mostly grown in Colombia and Ecuador. Since the U.S. requires that they must be free of pests and disease, copious amounts of pesticides and fungicides are applied to the flowers and plants, which poisons the farm workers and pollutes local water and wildlife.

Palm oil is increasingly used in many household products, including ice cream, crackers, detergents, and cosmetics. The production of soya and palm oil is done by clearing rainforests, destroying the habitat of many species of plants and animals. It also contributes to global warming by releasing large amount of carbon dioxide into the atmosphere. Demand for exotic foods distorts the local market in other ways. Quinoa is a nutritious grain that was a staple of the diets of Incas and their descendants for millennia. Its cultivation was discouraged by Spanish conquerors, but it remained a component of the local diet in the Andean region, primarily Bolivia, Peru, and Ecuador. Nutritionists in the Western world discovered about two decades ago that this grain is a rich source of protein that contains all essential amino acids and can be a replacement for animal proteins. The demand for quinoa in the U.S. and Europe has exploded and its price increased by more than a factor of three during the five-year period from 2008 to 2013, making it difficult for the indigenous population to afford it. Numerous varieties of quinoa are grown in the Andes region, varying in color, including white, yellow, gray, brown, pink, and red. While some types of quinoa are bland, others are bitter. The demand of Americans and Europeans for the bland, yellow variety puts selective pressure on farmers to grow only those grains, thus threatening the loss of genetic diversity.

Consequences of Overconsumption

There are a number of problems associated with the culture of consumerism that are becoming increasingly important with each passing year. The production of commodities requires natural resources such as minerals, fossil fuels, and water. Almost all production is now done with machines that require energy, usually produced with fossil fuels. The manufacturing facilities release toxic effluents that pollute and degrade the environment. As we continue at an ever-increasing pace to search for more and more amenities, the ecosystem must provide the resources. In this process, forests are cut down, oceans are denuded of aquatic life, land gets degraded due to the accumulation of harmful chemicals, air is polluted with smog and other substances, many species become extinct due to loss of habitat, and chemicals seep into the groundwater, thereby endangering human health. Ignoring environmental constraints and assuming that consumption-driven growth will continue forever is wishful thinking, but very convenient for those who are beneficiaries of the present system.

Although consumers are not always aware of it, the manufacturing process of various types of consumer goods—from the production of raw materials to their eventual disposal after use—has deleterious effects on the environment. Regarding the raw materials, petrochemical products pollute the air with the emission of gases, including greenhouse gases. Cotton farming requires large amounts of water, and chemicals applied in farms degrade air and water. Wood is harvested by cutting down forests that maintain the balance of air and water in the ecosystem. Substances that are used to produce manufactured goods often release harmful chemical effluents. Even the finished products that are sold in markets may have harmful health effects on users. BPA (bisphenol-A) and PBDE (polybrominated diphenyl ethers) are generally present in mattresses, pillows, rugs, curtains, carpets, padding, televisions, and computers.

PFOA (perfluorooctanoic acid) is used in Teflon pots, rugs, couches, and stain-resistant waterproof items. PFOA and PFOS (S-sulfonate) are virtually indestructible. BPA is ubiquitous because it is used in making plastics. Studies have linked these chemicals to a host of ailments including cancer, sexual problems, and behavioral issues.[244] They also accumulate in the ecosystem so that their effects persist for a long time. Even a personal computer contains numerous chemicals that are harmful for human health, including chromium, mercury, beryllium, lead, cadmium, and barium.

Factories that produce manufactured goods invariably release toxic chemicals that contaminate air and water. Since much of this pollution is unacceptable by American standards, many factories owned by multinational corporations are located in developing countries so that the environmental degradation takes place in distant lands. According to the World Bank, sixteen of the world's twenty cities with the worst air are in China. Outdoor air pollution contributed to 1.2 million premature deaths in that country in 2010;[245] the numbers for India are somewhat similar. Also, a fifth of China's farmland is polluted due to effluents from factories. Thirty percent of the electricity in the U.S. and 79 percent of the electricity in China is produced by coal-fired power plants. Generation of electricity from coal-fired power plants anywhere in the world contributes to climate change by emitting greenhouse gases. Since China is a major supplier of manufactured goods to the whole world, the greenhouse gas emission from that country is exceptionally large. It is useful to remember that pollution does not remain confined within narrow geographical boundaries and slowly degrades the environment everywhere.

Americans use about 100 billion cans per year for drinking beverages, which translates into about 340 cans per person per year. This is ten times more than the average number of cans used by Europeans. With less than 5 percent of the world's population, the U.S. uses one-third of the world's paper, a quarter of the world's oil, 23 percent of the coal, 27 percent of the aluminum, and 19 percent of the copper. David Tilford of the Sierra Club says: "Our per capita use of energy, metals and minerals, forest products, fish, grains, meat, and even fresh water dwarfs that of people living in other parts of world."[246]

Exploitation of Labor

With optimization of profit as the sole criterion, labor in developing countries is paid the very minimum for their monotonous, backbreaking, and often dangerous work. The workers are often forced to live in subhuman conditions. The pressure of globalization has led many companies to hire child labor. The cocoa industry hires 15,000 children in conditions of forced labor for picking beans in Ghana and Ivory Coast. About 218 million children between the ages of five and seventeen work in factories in various parts of the world, about 73 million of these children work in hazardous occupations.[247] Famous American brands use cheap labor in Southeast Asia to reduce the cost of labor and to avoid regulations on working conditions in Europe and the United States. In some cases, these companies have hired goons to intimidate workers and union leaders in those countries.[248]

Climate Change

Overconsumption almost always involves fossil fuels in one form or the other. A substantial amount of clothing and household furnishings are made with petrochemical products. The machines used to fabricate consumer goods either run directly on fossil fuels or use electricity produced from this source of energy. Most of the food that is grown on industrial farms depends on the use of fertilizers and pesticides made from fossil fuels. Transportation of goods to distant lands by ships or planes involves the combustion of fossil fuels, thus releasing even more carbon in the atmosphere. Luxurious living in outsize homes and driving exceptionally large vehicles also has an environmental cost because these things contribute to climate change.

Overconsumption and Clutter

Even though the average American home has tripled in size during the last fifty years, we fill them to capacity—garage, attic, and basement. Most people who have garages cannot park cars in them because of numerous items that cannot be accommodated in the house. The average U.S. household has 300,000 things. American children make up 3.7 percent of the children in the world but have 47 percent of the toys and other playthings.[249] According to a study in England, a woman in her lifetime will own 620 dresses, 434 pairs of shoes, 310 skirts, 588 pairs of trousers, and 372 cardigans.[250] Corresponding numbers for American women are going to be substantially greater.

Our numerous possessions greatly exceed the capacities of houses, even though the average American home is 1,000 square feet larger than in 1973.[251]

The continuous increase of possessions has made self-storage a roaring $24 billion business in the U.S. The storage capacity of these establishments is so large that every American could fit inside their units, simultaneously.[252] There are 48,500 storage facilities in the United States, compared with only 10,000 in the rest of the world. These facilities now outnumber all McDonald's, Wendy's, Burger King, and Starbucks outlets in the U.S., combined. Since most people find it difficult to part with their possessions, they have to take the help of professionals when they need to downsize.

In addition to seductive advertising and the availability of easy credit, there are other reasons for continuously increasing possessions. Most consumers are unaware of the environmental impact of articles that are kept in attractive packages in the showroom. Some of them may even contain toxic materials or may have been produced in a manner that pollutes the local environment, while others may have been produced by workers without any rights or under exploitative conditions. It is also a common belief that technological developments will find a solution to any problems that we create for the present and future generations. This misplaced faith in technology ignores the fact that the problems that we are creating, such as degradation of the environment, depletion of resources, and emptying the oceans of marine life are global problems that cannot be solved by the invention of new gadgets.

Mountains of Trash

A serious problem created by the culture of consumption is the amount of waste produced by homes, businesses, and industrial establishments. The discarded garbage takes precious space, emits noxious gases into the atmosphere, and releases toxic chemicals that pollute the groundwater. Municipal Solid Waste (MSW) is garbage that is primarily a product of the extravagant lifestyle and throwaway culture. Accumulated waste is a serious problem in all countries. The rate at which waste is being produced far exceeds the capacity of the ecosystem to absorb or recycle it. The World Bank estimates that the 3 billion people living in urban areas generate 1.4 pounds of MSW per person per day. By 2025, it is projected that 4.3 billion urban residents will produce a worldwide total MSW of 4.8 trillion pounds (2.2 billion metric tons) per year. The staggering amount of solid waste produced each year will eventually interfere with human lives on the planet.

Generation of vast amounts of solid waste is a relatively recent phenomenon. About a century ago, nearly all products used in daily life were fabricated from natural materials. Articles made of wood would eventually degrade through biological processes and metallic objects could be processed over and over again. The use of synthetic materials, those which are not amenable to biodegradation, has greatly increased in the last few decades, making recycling of waste through natural processes very difficult. Americans produced 258 million metric tons of solid waste in 2017, which amounts to 4.4 pounds of waste produced by each person per day. Home to only 4 percent of the population, Americans are responsible for more than 30 percent of the waste generated on the planet. The breakdown of municipal waste produced in the U.S. into categories is as follows:[253]

Municipal Waste	Percentage
Paper	32.7%
Yard waste	12.8%
Food waste	12.3%
Plastics	12.1%
Metals	8.2%
Rubber, metal, textiles	7.6%
Wood	5.6%
Glass	5.3%
Other	3.2%

The production of solid waste is increasing each year. While waste from households includes food waste, packing materials, containers, batteries, leftover pesticides, and various cleaning products, the waste from manufacturing facilities, automotive garages, and construction sites may contain a plethora of harmful chemicals. MSW is also increasing in toxicity as industries find new chemicals that improve the appeal of their products, without considering their effects on the environment or human health. At present, thousands of untested chemicals are added to consumer products each year.[254] American women apply, on average, 168 chemicals in the form of cosmetics, perfumes, personal care products, and feminine hygiene to their faces and bodies every day.[255]

The EPA generally takes action against a chemical only when its harmful effects have been well established.

Food waste in the garbage leads to the growth of pathogens and vermin that also invade neighboring homes and other areas. The food scraps in land-fills are broken down by microorganisms through fermentation, which releases methane, a greenhouse gas that is many times more potent than carbon diox-ide. Methane from landfill sites accounts for 12 percent of atmospheric meth-ane emissions and 5 percent of the total greenhouse gas emissions. Disposal of garbage in a pit in the ground, or even above ground, is practiced in much of the developing world. In the United States, the Resource Conservation and Recovery Act mandated that new landfills should be lined with plastic, clay, or both. They are also supposed to collect liquid or gaseous materials exuding from landfills. Technologically advanced landfills are expensive to design and operate and have replaced many old hole-in-the-ground disposal sites. Since a considerable investment is required in designing and operating these landfills, these mega landfills are fewer in number. In 1986, there were 7,683 dumps in the United States, while the number of newer landfills now is only 1,908.

The high cost of these facilities has resulted in the construction of mega facilities that use economy of scale. These large landfills are generally located in sparsely populated regions. The downside of this improvement is that since these modern facilities are fewer in number than older landfills, trash has to be transported over long distances, resulting in the emission of substantial amounts of greenhouse gases. At present, the biggest landfill in the U.S. is located in Apex, Las Vegas, Nevada, which receives about 4 million tons of trash every year. Municipal waste from New York City, which used to be dumped on Staten Island, is now shipped to Ohio, Pennsylvania, or West Virginia. Send-ing all the trash generated in New York City to West Virginia would generate 760,000 tons of carbon dioxide each year.[256] The waste industry is a big and profitable business that applies pressure on local municipal officials to send waste to landfills instead of putting greater effort into recycling.

Finding suitable space for creating landfills is a serious problem in most countries. Landfills in many states in the U.S. are running out of space and many existing landfills will be filled to capacity in a few years. Massachusetts and Rhode Island have just twelve years of remaining capacity. New York State, even though it sends much of its trash to other states, has only twenty-five years of capacity left until the existing landfills are filled to capacity. Waste disposal is a problem almost everywhere—the U.K. may run out of landfill sites in eight years. Construction of new landfills faces opposition from local residents due to environmental concerns. Modern mega landfills do not completely eliminate the problems of pollution caused by waste because some toxic materials invariably leach into the ground or are emitted as gaseous substances. The space that becomes uninhabitable due to the construction of landfills extends far beyond their physical boundaries because these landfills emit noxious gases into the atmosphere and become breeding grounds for dangerous and invasive species of animals and microorganisms. When water from rainfall percolates through the landfill, it picks up some materials in the waste such as corrosive chemical agents, disinfectants, and unused prescription drugs, as well as solvents such as paint thinners, pesticides, and heavy metals. This contaminated water, called leachate, escapes the landfill and mixes with the groundwater if the base of the landfill is even slightly permeable, which is usually the case.

For a few decades, wealthy countries of the Western world have been shipping their waste to countries in Asia, with the lure that indigent people in those countries can extract useful materials from the trash dumped on their shores. China stopped accepting such waste materials in 2018 and ships began unloading household garbage in other Asian countries, including Vietnam, Thailand, Malaysia, and the Philippines. However, these countries are realizing that waste materials keep accumulating at a rapid rate, polluting the local environment, and the advantages are insignificant. Therefore, many countries are imposing a ban on the importation of waste from North America or Europe.

We are accumulating billions of tons of junk that we have deposited, and continue to deposit, in our landfills. Due to its contents, accumulated waste is a health and environmental hazard that will persist in the ground for a very long time. It will create problems for future generations that we cannot comprehend. Incineration of waste releases dangerous chemicals into the air. Heavy metals and chemicals that do not degrade will remain in the ground for at least a few decades. The only solution to problems created by waste is to reduce its volume and toxicity.

Food Waste

Every year in the United States roughly 60 million tons (133 billion pounds) of food, estimated to cost about $160 billion, is discarded or thrown away.[257] An American family of four throws away about $1,484 worth of food each year.[258] The fermentation of food waste in landfills and dumps comprise18 percent of the total methane emissions in the country, which contribute to climate change. Food waste is a worldwide problem. Roughly one-third of the food produced in the world for human consumption—approximately 1.3 billion metric tons, worth nearly $1 trillion at retail prices—is lost or wasted.[259] This includes 45 percent of fruits and vegetables, 35 percent of fish and seafood, 30 percent of cereal, 20 percent of dairy products, and 20 percent of meats.[260] Although roughly the same proportion of food is wasted everywhere, the details are different in developed and developing countries. In affluent countries, consumers buy much more food than they will eventually consume and the excess food, including items that have lost their appeal or are past their expiry dates, is thrown away. In the developing world, a substantial amount of food is lost in storage and transportation, and a much smaller amount is thrown away by the consumers. In a world in which more than 870 million people do not get enough food on a daily basis, such an enormous loss of food is unconscionable. Just one-fourth of the food currently wasted would be enough to provide food to all hungry people in the world.

When food is not used for human consumption, all the resources that were used in producing it—water, land, labor, and agricultural inputs—are also wasted. Since the amount of food produced is often limited by the available resources, squandering food has a direct relation to the shortage of food at some other place. The energy that goes into the production, harvesting, transporting, and packaging of the wasted food generates more than 3.3 billion metric tons of carbon dioxide.[261] When food ends up in landfills, the emission of methane during the fermentation process adds to the accumulation of greenhouse gases. "If food waste were a nation, it would be the world's third largest emitter of greenhouse gases."[262]

Particularly in affluent countries, and to some extent everywhere, the appearance of fruits and vegetables is an important consideration for consumers. At the earliest stage of harvesting, ugly fruits are left to rot or sent directly to landfills because those items are not going to sell. Most supermarkets keep a large pile of fruits in an attractive manner and frequently cull them to remove items that have begun to deteriorate. Grocers also find it easier and more profitable to throw away fruits that have lost their appeal rather than to lower the price before that stage is reached. Manipulating the expiry date also increases sales. If customers throw away items because they are past their expiry dates, they will buy more of those items as replacements.

In a world where hunger is common and shortage of food often causes social unrest, wastage of food is completely unacceptable and represents a serious problem that must be addressed to provide food to more people, and to prevent environmental degradation by this sector. In rich countries, where consumers buy excessive amounts of food and throw away a substantial amount, better education of consumers will be helpful. Since the food losses in developing countries are mainly on the upstream end, governments may help by providing better storage and transportation facilities. The two crucial inputs to produce food—water and land—are already in short supply in many parts of the world. It is not possible to increase them to feed the present population and the millions more that will arrive by the middle of the century. Decreasing food waste will provide enough food to everyone for a very long time while the population stabilizes at an acceptable level. An extra dividend would be a decrease in the emission of greenhouse gases.

E-Waste

Electronic waste, or e-waste, from discarded electronic items such as computers, printers, televisions, and cell phones forms a separate category. These articles often contain harmful materials such as lead, cadmium, chromium, brominated flame retardants, and polychlorinated biphenyls (PCBs). Americans own approximately 24 electronic products per household. According to the EPA, 3.14 million tons of e-waste was generated in the U.S. in 2013, of which about 40 percent was recycled, 40 percent was exported, and the rest got dumped in landfills. The worldwide generation of e-waste is estimated to be 20 to 50 million tons per year. A very large fraction of e-waste is shipped to China, where workers remove valuable parts for reselling. According to a UN report, the town Guiyu in southern China has a large-scale industry to extract valuable components.[263] Researchers have linked e-waste to adverse effects on human health, such as inflammation and oxidative stress that are precursors to cardiovascular disease, DNA damage, and possibly cancer.[264] A United Nations convention has now imposed a ban on the transfer of electronic waste from developed countries to countries like China and Vietnam. However, a large amount of e-waste still ends up in China through illegal channels that first ship it to Hong Kong and then to China. It is creating an environmental calamity because circuit boards are burned and parts treated with hydrochloric acid to recover valuable metals. It is dangerous for the health of workers who extract usable parts from the circuit boards and other materials.

Other Sources of Pollution

Although solid waste from municipalities has received a lot of attention, two much bigger sources of solid waste in the U.S. are the livestock industry and the extraction and processing of minerals. A livestock farm, which houses hundreds of thousands of pigs, chickens, and cattle, produces waste at a scale comparable to that of a small city. The waste generated by these operations pollutes the groundwater, emits toxic and greenhouse gases, and gives rise to pathogens that cause diseases in humans. The widespread use of antibiotics in this industry contributes to the growth of resistant strains, making it harder to treat human illnesses. Large hog farms emit hydrogen sulfide gas, which can cause brain damage in humans after long exposure. Unlike municipal waste that is primarily produced in urban areas, livestock operations are set up in rural areas with low population density and hence degrade pristine lands. The proclivity to consume animal-based foods extracts an enormous price from the environment and also consumes precious resources.[265] The consumption of meats is rapidly increasing in most developing countries, particularly China. Mining and processing of minerals is another major source that pollutes air, land, and groundwater. Particulate matter released during the processing of metallic or non-metallic ores contaminate the groundwater during the process of washing minerals. The land in the vicinity of these plants becomes so polluted with the effluents from these facilities that it cannot sustain any form of plant or animal life.

Global Ecological Footprint

Overconsumption greatly increases the burden of humanity on the ecological resources of the planet. This load was quantified by Mathis Wackernagel and William Rees in the early 1990s with the concept of a global ecological footprint calculated from natural endowments used by people in various activities.[266] In the ultimate analysis, food, energy, and all amenities of life are produced from natural resources. The footprint is calculated by estimating the contribution of natural products provided by the ecosphere for the survival and welfare of humanity, including water used directly or indirectly, land used for agriculture, forests, and all types of minerals. It also takes into account the land area required for the assimilation of waste materials. Such considerations show that, at the present time, humanity uses the equivalent of 1.6 Earths to provide the resources and absorb the waste.[267] This number is greater than 1.0 because we are using the inbuilt resources of the planet and producing waste that is not fully assimilated in the ecosystem.

A numerical estimate of the extent to which humans are overexploiting the planetary resources is calculated by the Global Footprint Network as the "Earth Overshoot Day," the day of the calendar year when humanity has used up the resources produced by the Earth in that year and also exhausted the waste assimilating capacity of the planet. This day is calculated by using 6,000 data points per country for roughly 200 countries and then aggregating them into a single number. In 1970, when the human population was about 4 billion, the Earth Overshoot Day was in late December. Since then, it has been steadily moving up because we are using increasingly greater amounts of the natural resources of the planet with each passing year. It was July 29 in the year 2019.[268] Due to the slowdown of economic and other activities, it moved backward to August 22 in the year 2020. This is the first time that the total consumption of planetary resources has been less than in the previous year. If we consider 2019 to be a typical year, this means that people worldwide exhaust the Earth's capacity to support us in a sustainable way in less than eight months. The costs of ecological overspending are evident every day in the form of deforestation, expanding deserts, biodiversity loss, continuously increasing carbon dioxide in the atmosphere, acidification of oceans, loss of marine life, accumulation of mountains of waste, and air and water pollution.

Since ecological footprints measure the burden imposed by a people on natural planetary resources, it differs by large amounts between residents of different countries and also for people living in the same country. The per capita ecological footprint of selected countries, calculated in global hectares, is shown in Figure 8. The global hectare is a useful measure of the capacity of Earth to support people. It is obtained by converting all requirements of humans to a physical area and is used to determine the carrying capacity of our planet. The global hectare will be more for a lush forest than for a desert. The total capacity of the Earth has been calculated to be about 13.4 billion global hectares. The total ecological footprint of ten countries that have the greatest impact on the planetary resources and environment are as given below.[269]

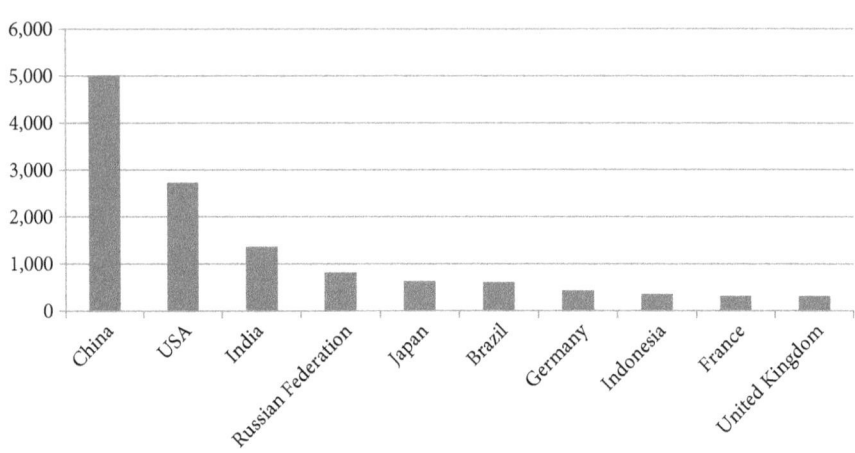

Figure 8 Total Ecological Footprint (global hectares).
Source: Global Footprint Network.

Calculating the per capita footprint of a country is also possible by dividing the total footprint by its population. The per capita global ecological footprint of a few countries is shown in Figure 9. It is highest for an American, indicating that an American lifestyle presents a greater burden on the ecological resources than most other lifestyles. Although China had the highest total ecological footprint because of its large population, its per capita contribution is rather modest. Of the countries considered for this analysis, the per capita footprint of an Indian is small.[270] Since many consumer products are obtained from other parts of the world, national boundaries are not important. The number for per capita footprint indicates the amount of land used by an average resident of the country to produce articles needed for daily living and to assimilate the waste generated by him. The fact that the total ecological footprint is 1.6 times the land area of the Earth indicates that we are drawing down on natural resources faster than they can be regenerated. Cutting down forests, depleting sources of freshwater, polluting land, air, and water, and overfishing that will affect the survival of marine species are some of the things that draw down from the global reserve. Most developed countries have an impact on the environment that is much greater than the resources within their geographical boundaries. Countries with a large footprint are usually dependent on natural resources from other parts of the world.

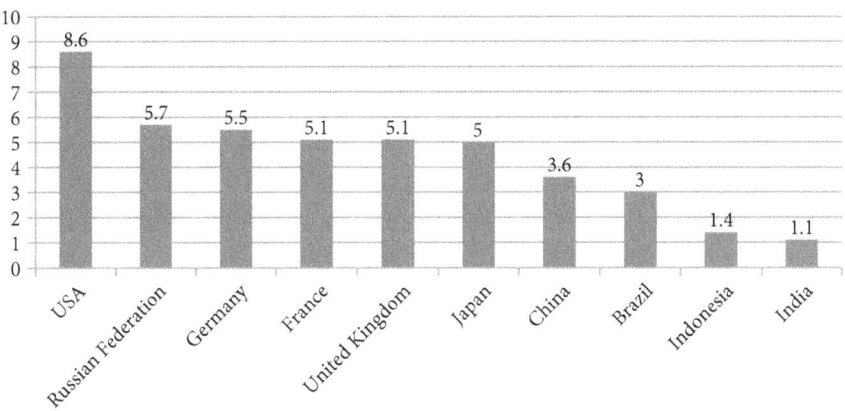

Figure 9 Per Capita Footprint (global hectares).
Source: Global Footprint Network.

It is obvious that humanity cannot keep depleting the ecosystem of stored products forever—at some stage the resource consumption, environmental degradation, and waste generation will begin to affect the life and lifestyle of all people, although not to the same extent everywhere. Sustainability requires that humanity's ecological footprint be reduced to within the limits of the planet. Economists and business interests that advocate increasing consumption forever ignore the constraints imposed by our finite planet. Increasing consumption to the level of Americans in all parts of the world is simply impossible because the planetary constraints will make life miserable before that happens.

Sustainable Consumption

A concept that is being discussed by national and international bodies it that of sustainable consumption, which requires that the use of ecological resources be reduced to the level of production of these items, and the generation of waste to also be kept within these bounds. In September 2015, the United Nations set seventeen goals for the sustainable development of the world. One of these goals was to ensure sustainable consumption and production patterns to reduce the use of resources by educating consumers and increasing their awareness of overconsumption.[271] This approach is also concerned with the health and welfare of workers who produce consumer goods. This could be achieved by regulations and taxes imposed by governments in the manufacturing countries, and awareness in the public. Products may have a "Fair trade" or "Ecolabel," which certifies that the product or service has met strict environmental criteria. This label will certify that the workers who made these products were given a living wage, had appropriate working conditions, and were not exposed to unhealthy chemicals. Although ecolabeling is a good idea because it provides consumers with socially and environmentally friendly choices, it fails to address the basic problem of overconsumption and may even lead to overconsumption of products that have the proper certification.

The Way Forward

In much of the world, in both developing and developed countries, increased consumption is considered a sign of higher status. It is promoted by sophisticated advertising and facilitated by easy credit and other means. Consumers are often unaware of the exploitive situation in the production of items that they buy, the burden of these products on the biosphere, and the long-term environmental effect of consumer goods. To achieve even a limited amount of sustainability, we have to stop buying things we don't really need—things that only end up in overstuffed closets, attics, or storage facilities. Often it is difficult to determine the carbon footprint of items, although, in general, items imported from across continents will have a larger carbon footprint than those produced locally. It would be highly beneficial if the level of consumption in the Western world was decreased and the growth of consumption in the developing world was arrested. It would be unconscionable for people in wealthy countries to demand that developing countries keep their consumption at a low level while they continue in their profligate ways. While some people in developing countries cannot even afford basic necessities, the newly emerging middle class in these countries tries to emulate the lifestyle of people in Western countries as best as possible, creating a mini-America in their midst.

It is important to begin with the realization—to be inculcated everywhere—that humans are part of nature and not above it. The ecosystem will not support continuously increasing consumption forever. The demand for decreasing consumption does not mean that we have to go back to the Stone Age, only that we have to dial it back by about two or three decades. When there is an urge to acquire something, instead of considering how useful that item will be, it is better to consider how deprived life will be without that object. It is also good to remember that the goal of all advertising—direct or indirect—is to sell a product and enrich the corporations and their agents. The joy of acquiring a new object is temporary. Increased consumption does not make us happier; on the contrary, the urge to acquire items that are beyond reach makes some people depressed.

SIX

· ·

ANIMAL-BASED FOODS AND SUSTAINABILITY

The amount of food produced by the ecosystem depends on the available resources, such as suitable land for farming, fresh water, and climatic conditions that are conducive to the growth of plants. The factors that have made the current food situation very precarious are the human population of 7.8 billion, increasing at the rate of about 80 million per year, and changing food preferences in favor of animal-based foods—various kind of meats, eggs, and milk. The conversion of agricultural products to animal-based foods involves a large wastage of energy because part of the food eaten by animals sustains them during their lives and develops body parts not fit for human consumption. While the demand for animal products is increasing, the capacity of the land to produce animal-based foods is limited by the availability of resources in many regions of the world. In addition, the production and consumption of animal-based foods has deleterious effects on the environment and human health. These factors make it necessary to consider the foods that we choose to eat and their effects on the ecosystem and on the capacity of the planet to keep providing this basic human requirement.

Animal-Based Foods: Meats

Foods derived from animals constitute a major component of the diet in developed countries and their consumption is rapidly increasing in middle-income countries, particularly China. According to data published by the U.S. Department of Agriculture, an average American ate 222.2 pounds of red meat and poultry and 146 pounds of milk products in 2018. The United States leads the world in the per capita consumption of meat, which is produced from the slaughter of 25 million farm animals each day. The average meat consumption in most developed countries is between 150 and 200 pounds per year. While the per capita meat consumption in developed countries has somewhat stabilized at a high level, it is rapidly increasing in China. Although the per capita consumption of meats in that country is lower than in most Western countries, the population of about 1.4 billion and rapidly increasing consumption makes China the world's largest consumer and importer of meats. Brazil is a major producer and exporter of meats, particularly beef.[272] Over the past fifty years, the worldwide production of meat has more than quadrupled while the population doubled because the per capita consumption has greatly increased in the world. The number of farm animals slaughtered each year to meet the demand is very large—about 50 billion chickens, 1.5 billion pigs, 550 million sheep, 444 million goats, and 300 million cattle ended up on dinner tables in 2015.[273] The mix of animals is somewhat different in each region, but the number of farm animals killed for food is very large almost everywhere. According to the United States Department of Agriculture, 8 billion chickens, 277 million turkeys, 36 million cattle, and 124 million pigs were slaughtered in the country in 2018.

The production of such large amounts of animal-based foods uses precious planetary resources and creates many environmental problems. Since meat and milk are secondary products of the agricultural industry, it takes more planetary resources to produce these foods for humans than the direct consumption of agricultural products in a vegan lifestyle. The conversion of agricultural products into meat, milk, or eggs involves wastage of energy and nutrients, and part of the feed consumed by farm animals sustains them until they reach the desired weight or begin to give milk or eggs. The production of animal-based foods requires more resources, such as land, water, fertilizers, and other chemicals, than the direct consumption of agricultural products by humans. The food requirements of the burgeoning human population have already stretched the use of land and water to the limit. There is an acute shortage of water at many places where the productivity of agricultural lands is limited by the availability of water. The only way that additional land can be brought into cultivation is by razing forests that play an important role in preserving the ecosystem, including producing oxygen for the benefit of all living things. Other benefits of forests include providing habitat for a great variety of animals, protecting watersheds, preventing soil erosion, and mitigating climate change.

Since the production of meats and dairy products is an intrinsically inefficient process, a meat-based diet greatly increases the requirements of agricultural products. Farm animals also pollute and degrade the land, emit noxious and harmful gases that also contribute to global warming, and cause ecological problems that will interfere with the welfare of humanity. The production of animal-based foods requires an order of magnitude more resources than the direct consumption of equivalent plant-based foods by humans. The world produces 350 million tons of meat each year.[274] Plant-based replacements for each of the major animal categories in the United States (beef, pork, dairy, poultry, and eggs) can produce two- to twenty-fold more nutritionally-similar food per unit of cropland. Replacing all animal-based items with plant-based foods will save enough food to feed 350 million additional people in the country.[275]

Producing animal-based foods presents a huge burden on the ecosphere. The 70 billion farm animals in the United States consume 70 percent of the grain grown in the country. In addition, about 788 million acres, 41 percent of the land excluding Alaska, are used for grazing by livestock in the early part of their lives. The fraction of land devoted to the production of animal-based foods is large everywhere. According to the FAO, 26 percent of the Earth's terrestrial surface is used for grazing by livestock and one-third of the arable farmlands are used to grow feed for them. The huge livestock industry uses precious planetary resources and causes environmental degradation in most regions of the world. To meet the demands of the livestock industry, forests are razed and converted into farms for growing feed or to be used as pastures. The livestock industry rears and slaughters 83 billion animals around the world each year (more than 220,000 per day) which has an enormous detrimental impact on the ecosphere. About 25,000 farm animals are killed for food each day in the United States.

Animal Factories

To satisfy the demand for meat and milk, the number of cattle and dairy cows raised at any time is very large—about 94.4 million animals are kept in facilities in all fifty states of the U.S. at any time. Privately owned range- and pasturelands cover 27 percent of the total area of the forty-eight contiguous states, exceeding both the area covered by forests (21 percent) and farmlands that grow crops (18 percent).[276] The large amount of animal products consumed by the human population has been made possible by the development of animal factories known as concentrated animal feeding operations (CAFO). Almost 99 percent of the animals in the United States are raised in such facilities and about 90 percent in the rest of the world. Raising hundreds or thousands of animals in a large CAFOs reduces the per-head cost of raising animals. Major companies that own these facilities improve efficiency and reduce expenses by treating animals as mechanical objects. A typical CAFO in the United States contains 1,000 beef cattle, 700 dairy cows, 25,000 swine, 125,000 boiler chickens, or 82,000 laying hens, but there could be significant variations in the capacities of these units. There are approximately 10,000 CAFOs in the U.S., each keeping one type of animal.

Livestock spend the first part of their lives grazing on pastures or rangelands. Because of the large number of farm animals, more than one-third of the land in the U.S. (654 million acres) is used for pasture and another 127.4 million acres is used to grow feed for livestock.[277] Between pastures and croplands used to produce feed, 41 percent of U.S. land in the contiguous states is used by the livestock sector in the country. The high concentration of cattle on these lands interferes with the habitation of other species that are native to those regions. The USDA Wildlife Service eliminates thousands of wolves, bears, cougars, and coyotes on rangelands each year to protect livestock. Cattle cause much damage to the ecosystem that decreases the productivity of rangelands. They eat all grasses and plants, leaving only unpalatable weeds. The ground gets compacted by their hooves, which makes it harder for the land to absorb water when it rains and difficult to recover to the productive state. Rivers and streams get polluted with excrement. The deposition of manure in the rivers and stream in these areas pollutes the water and even obstructs the flow of streams. By destroying vegetation, damaging wildlife, and disrupting natural processes, livestock cause significant harm to rangelands.

When cattle reach the rapid-growth stage, they are moved to feedlots and given a diet of corn and other grains that makes them gain weight rapidly and reach the desired weight for slaughter. While cattle spend the last four to six months in feedlots until they acquire a weight of about 2,000 pounds, pigs spend their entire lives of five to six months in such facilities and are slaughtered at the finish weight of about 250 pounds. Although the physical footprint of these CAFOs is not large, they are supported by huge tracts of farmland to provide feed for the animals. Pigs and lambs are also given a diet of cereal grains in feedlots. Since a diet of cereal grains is not suitable for the digestive system of farm animals, they are given a daily dose of antibiotics. Another reason for giving these drugs to them is that in the unnatural confined spaces of feedlots, they suffer from greater stress and risk of succumbing to disease. However, the most important reason is that antibiotics act as growth promoters and animals grow faster when given these drugs. According to the Food and Drug Administration, 80 percent of the total antibiotics used in the country are given to farm animals. The use of antibiotics for increasing the weight of animals has been banned in the EU and some other countries. Excessive use of these drugs gives rise to strains of bacteria that are drug resistant and a person infected by these superbugs cannot be treated with modern medicines.[278] The widespread use of antibiotics is contributing to the public health crisis of antibiotic resistance. When superbugs reach humans through the food chain, they pose a grave danger to health. Each year, at least 2.8 million U.S. residents suffer from infectious diseases caused by drug-resistant bacteria, resulting in about 162,000 deaths.[279]

CAFOs cause air pollution through the emission of noxious gases such as ammonia and hydrogen sulfide, and gases from manure. Decomposition of waste releases harmful gases and airborne particles from the dry manure and animal dander, which can be transported by air for miles. The air in and around these facilities also contains some of the antibiotics given to animals. Livestock also produce an enormous amount of liquid and solid waste. These facilities produce roughly thirteen times the total waste produced by the human population in the country each year. Animal waste releases numerous harmful substances into the atmosphere, including ammonia, hydrogen sulfide, endotoxins, particulate matter, and airborne antibiotics. Since a CAFO typically holds a large number of animals, the high concentration of these substances creates a health hazard for the community. The waste from animal factories is accumulated in large ponds called lagoons. In addition to animal waste, these lagoons also contain antibiotic residues, cleaning solutions, and other chemicals used in these facilities. Each CAFO is surrounded by a few such large foul-smelling ponds. The nutrient-rich content of lagoons is conducive to the growth of pathogens, including bacteria, viruses, and other dangerous organisms that eventually contaminate neighboring land and bodies of water. The volume of the accumulated waste is so large that its disposal creates a problem for the livestock industry. Unlike human waste, the content of lagoons is not processed as sewage and is applied to agricultural farms without any treatment. The common practice is to apply it to farms in the neighborhood as a fertilizer, even when the farmland may be saturated with manure and cannot productively use the additional amount. When the untreated waste is applied to farms that have harvest-ready produce, it also gets contaminated by dangerous pathogens that may eventually reach the food bought by consumers. The liquid waste from animal factories flows into rivers and streams and eventually ends up in coastal seas, contributing to the creation of dead zones that cannot sustain any form of marine life. One of the largest dead zones is in the Gulf of Mexico, where nutrient-laden water from the Mississippi River flows into the sea. CAFOs are dusty places where bacteria are present in the air and transported to distant regions with dust particles and water droplets. Numerous studies have found high levels of antibiotics and antibiotic-resistant bacteria in the air samples downwind of feedlots. Flies and other insects that thrive around CAFOs can spread resistant disease into neighboring communities. The air near facilities that keep pigs, chickens, and cattle in large numbers not only stinks but can also be hazardous to public health.

Factory farming gets the blessing and support of international organizations such as the World Bank's International Finance Corporation (IFC), which has invested more than $1.8 billion in major livestock and factory farming operations around the world. This organization has financed the expansion of major multinational meat and dairy industries in Asia, Africa, Eastern Europe, and Latin America to produce dairy products and various types of meats.[280] The IFC says that this investment will create jobs and reduce poverty, but critics contend they harm the environment and only a few rich owners benefit from these grants.

Water Use and Pollution in Factory Farms

Unlike in earlier times, the availability of fresh water cannot be taken for granted in many parts of the world. Water scarcity is among the main problems to be faced by many societies and affects regions on every continent. Around 1.2 billion people, or almost one-fifth the world's population, lives in areas of water scarcity and another 500 million people are approaching this situation. Water use has been growing at more than twice the rate of the increase in population due to lifestyle changes and the consumption of animal-based foods. Although there is no global water scarcity as such, an increasing number of regions are chronically short of water. There is enough fresh water on the planet for the entire human population, but it is unevenly distributed and too much of it is wasted, polluted, and unsustainably managed.

Livestock operations require large amounts of water in all phases of operations. Water is required for growing feed, just as water is required to grow all agricultural products. Since one-third of the world's grain and about 80 percent of the soy is fed to farm animals, water used to grow these crops has to be attributed to the livestock industry. Rangelands often do not have enough natural sources of water to quench the thirst of animals, hence it is provided to them from other places. When animals are moved to feedlots, the need for water increases because of the diet of grains given to them. A large amount of polluted water is produced in these facilities from things like washing away the animal waste and servicing animals, which is collected in lagoons. The lagoons associated with CAFOs contain a toxic mix of solids and liquids produced in the facilities. Finally, copious amounts of water are used in abattoirs to wash the carcasses and prepare them for shipping to the market. In total, it takes five to ten times more water to produce animal-based foods than the direct consumption of agricultural products with equal nutritional contents.[281] According to the Water Footprint Network, it takes roughly 300 to 1,500 liters of water to produce 1 kilogram of most fruits and cereals, but between 4,000 and 15,000 liters for equivalent amount of animal products. An average American household uses about 5,700 liters (1,500 gallons) of water per person per day, almost half of which is associated with the consumption of meat and dairy product.

Since farm animals consume about a third of the agricultural produce in the world, water used for growing the feed items should be attributed to this industry. The water footprint of any product, defined as the water required to produce it, is much greater for animal-based foods than the water footprint of crop products with equivalent nutritional value.[282] The water required to produce some common food items is given below:

Water Requirements in Liters for 1 Kilogram of Each Item:[283]

Beef: 15,415
Sheep meat: 10,412
Pork: 5,988
Chicken meat: 4,352
Rice: 2,497
Wheat: 1,608
Apples: 820
Bananas: 780
Potatoes: 287
Cabbages: 257
Tomatoes: 214

Water Pollution

A somewhat related problem is that of pollution, which makes the water use-less even when it is available. Animal farming pollutes surface and groundwater by direct discharge and through seepage of pollutants. The common pollutants include bodily fluids of animals, chemicals used in the facilities, drug residues, pathogens, and microorganisms. Polluted surface water can kill fish, spread infections, and be unsuitable for human use. According to the EPA, excrement from factory farms has contaminated the groundwater in seventeen states and polluted 35,000 miles of rivers in twenty-two states.[284] Contents of lagoons can seep through the soil into nearby sources of groundwater and overflow during storms. In California, the nation's biggest dairy producing state, a study found animal agriculture to be responsible for serious nitrate contamination in areas with big dairy operations. The National Water Quality Inventory Report of 2002 noted that agricultural runoff was the leading cause of the pollution of rivers and streams, and the second leading cause of pollution in lakes and res-ervoirs. Runoff from factory farms causes algal blooms due to their high nutri-tion contents. When the algae die, they remove oxygen from the water, creating dead zones that cannot support any form of life. There are hundreds of dead zones in coastal seas around the world. One of the biggest dead zones is in the Gulf of Mexico where the Mississippi River discharges water from farms in the Midwestern states of the country.

The highly toxic content of lagoons pollutes the surface and groundwater by leakages from these ponds and also when it is spread on neighboring farms. Polluted water can kill fish, emit odors, and spread pathogens to the local pop-ulation. Runoff of synthetic fertilizers and animal waste can poison drinking water and aquatic biosystems, wreaking havoc on human health and killing wildlife. Bodies of water are also contaminated with fecal coliforms in the run-off from these facilities. Factory farms dump tens of millions of tons of animal waste and agricultural chemicals into the environment each year, polluting the land, water, and air of those regions. Growing the feed of farm animals also has deleterious effects on the environment. Over a billion pounds of pesticides are applied to crops in the U.S. each year, many of which persist in the soil for a long time. They are used to kill insects and weeds that damage the fecundity of farmlands, but they also poison the ecosystem. When they develop resistance to these chemicals, more potent pesticides must be used. The residues of these chemicals are found at every level in the food chain. Since a third of agricultural produce is grown for the livestock industry, the proportion of agrochemicals attributed to it is very large.

Drugs and Chemicals used in Animal Factories

According to the FDA, more than 20 million pounds of medically important antibiotic drugs were given to livestock in 2014—about 80 percent of all antibiotics used in the country.[285] Antibiotics are freely used in large amounts in some Asian and South American countries. Despite some restrictions on the use of antibiotics in animal factories in the U.S., low doses of antibiotics are still given through the feed to prevent the diseases that would result from the crowded and unsanitary conditions in factory farms. Because of cramped conditions, poor sanitation, and antibiotic overuse, dangerous bacteria may develop in these farms that could infect the human population through the environment or by ingesting animal-based foods. Trace elements, some of which are very toxic, are considered to be essential dietary components and given to almost all livestock species. Of particular concern in this regard is the feeding practice of pharmacological zinc and copper doses for the purpose of performance enhancement in pigs, poultry, and dairy cattle. Pigs excrete approximately 80 to 95 percent of the copper and zinc dietary supplements given to them, producing metal-enriched manure. When supplemented in doses above the normal requirements, these chemicals accumulate in the manure and present a potential threat to the environment. Adverse environmental effects of these elements include impairment of plant production and accumulation in edible animal products and local bodies of water.

Antibiotic Resistance

Frequent use of antibiotics may give rise to resistant strains by random mutations that are immune to the effect of these drugs. When antibiotic-resistant bacteria become widespread, lifesaving antibiotics become ineffective, forcing researchers to constantly develop new drugs. Antibiotic-resistant bacteria that often develop in factory farms pose a grave threat to human health. In recent years, the discovery of new antibiotics has slowed while strains of bacteria resistant to each new drug keep developing. The CDC estimates that each year in the U.S. at least 2.8 million people acquire infections from antibiotic-resistant strains of bacteria and 35,000 people die from the diseases caused by them.[286] A study by the *Review of Antibiotic Resistance* estimates that by 2050, if nothing is done to curb antibiotic misuse, resistant bacteria will kill 10 million people per year.[287] Antibiotic resistance is the third leading cause of death in the United States, behind heart disease and cancer. Since 80 percent of the antibiotics used in the country are given to farm animals, overuse of these drugs on livestock is the major reason for this impending calamity.

Climate Change

Perhaps the most pernicious effect of the livestock industry is that it produces large amounts of greenhouse gases that contribute to climate change. Animal factories produce greenhouse gases in almost all phases of operations. Because of the large number of farm animals raised to satisfy the demand for animal-based foods, one-third of the arable land in the world is used to grow feed for them. These farms use fertilizers and agricultural chemicals that are petrochemical products. The operation of the farms, feedlots, abattoirs, and the transportation of animals is done with machines that run on fossil fuels and emit carbon dioxide in the atmosphere. This gas is also emitted in all phases of meat production and storage. However, the greatest contribution to global warming and consequential climate change is made by the industry when the animals are in the feedlots. Cattle and sheep produce large amount of methane during the process of digestion, called enteric fermentation. Some methane is also emitted from the lagoons that contain animal waste. Methane is twenty-eight times more potent than carbon dioxide in its global warming effect. Nitrous oxide (N_2O), another greenhouse gas that has a 265-times greater warming potential than carbon dioxide, is emitted from the bodily fluids of farm animals in pastures and from lagoons in feedlots. In total, livestock produce 40 percent of the global methane emissions, 60 percent of the nitrous oxide emissions, and 9 percent of the carbon dioxide emissions, which makes the livestock industry responsible for 14.5 percent of the total global greenhouse gas emissions from all industrial and domestic operations in the world.[288] In addition to these factors, carbon dioxide is also released when forests are razed to make pastures or to grow feed for animals. Since forests act as sinks of carbon dioxide, deforestation increases the load of carbon dioxide in the atmosphere. For comparison, the total greenhouse gases emitted by the highly visible transportation sector (cars, trucks, planes) during regular operation is 14 percent of the total in the atmosphere. The copious amounts of greenhouse gases emitted by the livestock industry make a large contribution to global warming. The resultant higher temperatures are detrimental for the livestock industry in two ways—the output of agricultural farms decreases with increasing temperatures and the need for water by farm animals increases at higher ambient temperatures.

Biodiversity Loss

The livestock industry has an adverse effect on the ecosystem of rangelands in many ways—a staggering 60 percent of global biodiversity loss is caused by livestock populations encroaching on the habitat of wild animals and depriving them of resources necessary for their survival.[289] Livestock directly impact the survival of other species through grazing and trampling, changing vegetative cover and altering the natural habitats. Livestock farmers also eliminate wildlife that may pose a threat to their animals. Alterations in the natural flow of rivers and streams for the benefit of livestock eliminates marine animals in those bodies of water. Razing forests to convert those lands to pastures or to grow feed for livestock destroys the habitats of many species that lived there. Overstocking farm animals in rangelands also leads to land degradation and consequential loss of plant and animal life. The replacement of complex forest biosystems with pastures decreases biodiversity because the animals that used to live there lose their habitats. Nutrient pollution of the water of rivers and streams from the waste of animals causes eutrophication and acidification. A striking example is nutrient runoff from grazing systems in the eastern coast of Australia, which adversely affects the Great Barrier Reef. Fertilizers and other chemicals used in the agricultural farms of the Mississippi River decrease the population of marine organisms.

Animal Facilities and Human Health

CAFOs degrade the local environment by the emission of gaseous and particulate substances, and also pollute groundwater. These pollutants have an adverse effect on the health of workers in these facilities and people living in the neighborhood. Dozens of gases are emitted from the animal waste that are injurious to human health, including ammonia, hydrogen sulfide, methane, and nitrous oxide. Inhalation of particulate matter and suspended dust has been linked to asthma and bronchitis. Ammonia is rapidly absorbed by the body and causes problems in the respiratory system. Long-term exposure to hydrogen sulfide may cause dizziness, sinusitis, and irreversible brain damage. Epidemiological studies suggest that people living near these facilities are at an increased risk of developing neurobehavioral symptoms and respiratory illnesses. Livestock production is also an important contributor to water degradation when chemicals and pathogens from the waste seep into groundwater.

Some of the chemicals used in factory farms, such as nitrates, ammonia, and pesticides, end up in the groundwater. Spills and leaks from lagoons, which are quite common, cause bacterial contamination of groundwater and impart a foul smell to the air. Nitrates in drinking water decrease the capacity of blood to absorb oxygen, causing the blue baby syndrome and sometimes leading to death. Nitrates in water have also been linked to higher rates of stomach and esophageal cancer. CAFOs are the leading contributors to pollutants in lakes, rivers, and streams in neighboring regions. The nutrient-rich water causes algal blooms and hypoxia (low oxygen) in water that kills fish and seagrass and also reduces fish habitats. Globally, deforestation for animal grazing and growing feed crops is estimated to increase the load of carbon dioxide in the atmosphere by about 2.4 billion tons every year. Ventilators in the windows of these facilities can carry dangerous contaminants in air to distant places, affecting the health of people.

Promotion of Meats by Governments

The governments of many countries, including the U.S., Canada, and most European countries, promote the consumption of meats by giving subsidies to this industry. The American government spends $38 billion each year to subsidize meat and dairy, and only 0.04 percent of that amount to subsidize fruits and vegetables, even though federal guidelines urge people to eat more fruits and vegetables and lesser amounts of animal-based foods. Meat and dairy industries receive both direct and indirect support from governments. Federal subsidies have made corn and soy the cheapest feed for cattle; 20 pounds of feed grains produce one pound of meat. Payment of subsidies is skewed toward larger producers of these commodities. Copious amounts of water are required in all phases of the production of meat, which is also heavily subsidized. Calculations show that without subsidies for water and other things, hamburger could cost about $35 per pound. Every person in the country pays for these handouts to the livestock industry, whether they eat meat or not.

Health Effects of the Consumption of Meat

Meat is considered healthy because it contains substantial amounts of protein and most B vitamins. However, it does not contain vitamins A and C, and lacks dietary fibers. Meat also contains fairly large amounts of saturated fats that have many deleterious effects on human health. Since it is an important component of the diet in most countries, many studies have been carried out on the health effects of regular consumption of meat. A review of epidemiological studies in Europe and the U.S. showed that the consumption of red meat and processed meat—hot dogs, deli meats, sausages, and bacon—is associated with an increased risk of cardiovascular disease, type 2 diabetes, and total mortality in both men and women.[290] Only two servings of processed meat a week increased the risk of heart disease and stroke by 7 percent.

A large-scale prospective study of 37,000 men and 83,000 women was carried out by Dr. Frank Hu of the Harvard School of Public Health that followed the meat consumption and health incidents of these people over a period of about twenty-five years. The researchers found that one additional serving of unprocessed red meat raised the risk of total mortality by 13 percent. An extra serving of processed red meat such as bacon, hot dogs, sausages, and salami raised the risk by 20 percent.[291] An accumulated body of evidence established a clear link between a high intake of red and processed meat with a greater risk of heart disease, cancer, and diabetes that results in premature death. For processed meats, there is a much stronger association with a greater risk of heart disease and cancer, especially colon cancer. Processed meats contain high amounts of additives and chemicals that may contribute to the additional health risk. A study found that women who ate large amounts of red meat were more than twice as likely to suffer from hormone-related breast cancer, probably caused by the chemicals added during meat processing or the growth hormones given to cattle. In a related study, Eunyoung Cho and colleagues found that the consumption of red meat increases the risk of breast cancer in pre-menopausal women by a small amount.[292] Poultry can have the same impact on cholesterol as red meats. Most of the retail chickens are contaminated with intestinal bacteria, including E. coli, enterococcus, and salmonella. Chickens are often given arsenic, a highly poisonous material, to promote quicker weight gain. In a recent study, Joan Sabate and colleagues found that excessive meat consumption is associated with an increase in the risk of type 2 diabetes.[293]

Meat and Pandemics

Some of the most common and deadly human diseases are caused by bacteria or viruses of animal origin, such as avian flu, influenza, leprosy, MERS, and SARS-COV2. "Wet markets"—complexes of stalls selling fish, meat, and wild animals—are unique epicenters for transmitting potential viral pathogens because they provide an ideal opportunity for mutating pathogens to jump from one species to another, and occasionally to infect humans. It is thought that the stress of captivity and being brought to the market weakens the immune system of animals, which in turn creates an environment where mutating bacteria or viruses can jump from one species to another. When that happens, a new strain of virus may develop that can get a foothold in humans. The SARS-COV2 virus has already infected millions of people and taken the lives of an extremely large number of people in all parts of the world.

Animals in large-scale farms, oftentimes infected with diseases, provide a perfect breeding ground for pathogens from a weakened host. These farms keep animals tightly packed in high-population chambers with poor sanitation. They are generally given diets supplemented with antibiotics so that they do not succumb to disease. The antibiotic-rich animal waste pollutes waterways that often provide water for the irrigation of crops, thus contaminating the produce. Various types of bacteria and viruses also contaminate the meats sold to consumers. In fact, three-fourths of the dangerous diseases have emerged in the past from such zoonotic pathogens, including swine fever, mad cow disease, hantavirus, and avian influenza.[294]

The COVID-19 pandemic was expected to usher in the biggest retreat for global meat eating in a long time. Per capita consumption was expected to fall to its lowest level in a decade, representing the biggest decline since at least 2000, according to data from the United Nations. The pandemic has made people rethink their relationship with meat. The consumption of meats was predicted to fall to its lowest level in almost a decade in 2020. In the end, demand only declined by 0.5 percent, but is expected to sink still further around the world in every major market, according to the UN FAO's Food Outlook Report. This is being driven by economic hardship and post-pandemic concerns about where our food comes from, and the way animals enter the food chain. The reported origin of the virus in a meat market in Wuhan, China, and numerous COVID-19 outbreaks in abattoirs and meat processing plants around the world have had a major impact on the industry. Millions of animals—chickens, pigs, and cattle—will be depopulated because of the closure of processing facilities.[295] Although eating meat does not cause pandemics, the huge demand for meats and other animal products can only be met by factory farms or wet markets where these pathogens develop and proliferate.

Plant-Based Diet and Health

Plant-based foods contain many vitamins and minerals essential for a healthy immune system, such as vitamins A, C, and E, and trace amounts of zinc, selenium, and other minerals that are beneficial for health and cognitive functions. Whole foods offer some benefits over dietary supplements because they contain a variety of micronutrients and dietary fiber that reduce the risk of type 2 diabetes, colorectal cancer, and heart disease. Many fruits and vegetables are rich in antioxidants, compounds that fight free radicals and help counteract oxidative damages to tissues and cells. Studies have shown that plant-based diets consisting of whole grains, fruits, vegetables, and nuts are associated with a lower risk of cardiovascular disease.[296] Well-balanced vegetarian diets could prevent nutrient deficiencies and diet-related chronic diseases. Properly chosen vegetarian diets are better for health and for preventing diseases than meat-based diets.[297] However, plant-based diets consisting of fewer healthy plant foods, such as refined grains, potatoes, fried foods, and beverages high in added sugar, are linked to increased risk of cardiovascular disease.[298]

Food Waste—A Distraction

While discussing the environmental impact of food production, reducing food waste is often put forward as an attractive measure since it has the potential to save money for both producers and consumers. A halving of food waste at all stages will reduce emissions by 13 to 25 percent and land use by 11 percent. If food waste were a country, it would be the third largest emitter of greenhouse gases, behind China and the U.S. Upward of 40 percent of the food in the United States goes uneaten, which means all the resources used for producing that food are wasted. Although food waste is real, it is a distraction that is convenient to meat eaters. Food waste in developing countries is exceedingly difficult to solve because of the lack of facilities in farms and homes. Consumers in those countries are already very conscious; food is wasted due to lack of refrigeration and proper storage facilities. It will be nearly impossible to solve this problem in developing countries in the near future. Food waste in developed countries can be decreased with proper education but will continue until its price increases.

Environmental Footprint of the Egg Industry

An egg is considered healthy because it contains six grams of protein and important vitamins and minerals. In addition to direct consumption in various forms, eggs are used in many processed and baked foods. In 2017, about 319 million laying hens produced 92 billion eggs in the U.S.[299] It is estimated that a total of more than 1 trillion eggs are produced in the world each year. Most egg-laying hens, called layers, are raised in battery cages in which they are packed so tightly that there is no room to even spread the wings. The beaks of hens are usually trimmed to keep them from pecking and injuring each other. The feed given to them usually contains many additives and antibiotics to prevent the outbreak of disease in the crowded and dirty conditions of cages. While the lifespan of an average chicken is five to eight years, birds raised in factory farms are culled after two years, when their productivity decreases due to the stress of continuously laying eggs.

The air around the henhouses is polluted with ammonia, carbon dioxide, methane, and nitrous oxide. Ammonia is formed when uric acid in chicken manure breaks down. The dust on these farms contains high concentrations of microorganisms and particulate matter from the waste and feathers of birds that may obstruct the respiratory tract of exposed persons. Prolonged exposure to the polluted air may cause chronic obstructive pulmonary disease (COPD) and decreased lung function. Workers in these facilities face the greatest risk of these diseases, but the dust cloud emanating from these facilities also adversely affects the health of persons living nearby.

Milk and Dairy Cows

Millions of farmers worldwide tend about 270 million dairy cows to produce 909 million tons of milk each year. This industry impacts the environment in many ways because it is highly resource intensive. In the United States, cows drink 144 gallons of water to produce 1 gallon of milk. Nine percent of the cropland in the country (34.1 million acres) is used to grow feed for dairy cows. According to the

USDA, a farm with 200 dairy cows produces as much waste as community of 10,000 people. While the amount of milk produced in the U.S. has increased, the number of farms has decreased because of the consolidation of these facilities into megafarms that house 1,000 to 2,000 cows. The normal lifespan of cows is fifteen to twenty years, but continuous impregnation and milk production is very taxing for their bodies, and they are slaughtered at the age of about four or five years. In that time, their bodies become so flaccid that their meat is only good for making soups. The dairy industry has several deleterious effects on the environment and sustainability. The conversion of plant products to milk is an inefficient process, i.e., cows return only a small amount of nutrients fed to them as milk. Cows must drink a lot of water to produce milk—an important consideration due to the predicted shortage of fresh water in many parts of the world. Cows pollute the land with their waste, and methane produced in the digestive process contributes to global warming. A large number of animals kept in close quarters with their accumulated waste is conducive to the growth of insects and pathogens.

Vegan Junk Food and Sugar

Although a carefully selected vegan food can be as nutritious as animal-based foods and better for sustainability, the environment, and human health than a meat-based diet, it may be harmful if it does not contain enough variety to provide all nutrients and is loaded with nutritionally deficient items like oils and sugar. A junk food vegan diet, in which large proportions of calories come from carbohydrates and sugar, may be bad for the health and the environment.

Sugarcane is one of the major crops in all parts of the world and sugar is often the main component of unhealthy diets. Roughly 145 million tons of sugar is produced in 121 countries each year. The average American consumes almost 152 pounds of sugar each year, which equals 3 pounds (6 cups) of sugar each week. Sugar production takes a toll on the soil, water, and air, and may be responsible for more biodiversity loss than any other crop due to the destruction of native habitats. Growing sugarcane requires large amounts of water for irrigation and its production involves the use of many agricultural chemicals. The effluents from factories that make sugar from sugarcane contaminate neighboring bodies of water. In the United States, sugarcane farming has damaged the unique ecosystem of Florida's everglades, which has been converted from subtropical forests teeming with life into lifeless marshlands due to excessive fertilizer runoffs and agricultural chemicals.

Sugar is addictive and consuming sugary foods makes one crave more sugary foods. There is a difference between added sugar and sugars naturally present in fruits and some other agricultural products. The human body processes natural sugar slowly, which does not cause a spike of its level in the body, as happens in the case of added sugar. Consuming foods with added sugar can raise blood pressure and increase chronic inflammation, both of which may lead to cardiovascular diseases. Studies have found an association between a high-sugar diet and a greater risk of dying from heart disease.[300] Over the course of a fifteen-year study, people who got 17 percent of their calories from added sugar had a 38 percent higher risk of dying from such diseases compared with those who consumed less than 8 percent of their calories from added sugar. The study found that individuals who consume higher amounts of sugar-sweetened beverages tend to gain more weight and have a higher risk of obesity, type 2 diabetes, hypertension, and cardiovascular disease. Drastically reducing the amount of added sugar items in food would be good for the health of people and also reduce the environmental damages caused by sugar plantations. It is worthwhile to point out that excessive sugar consumption is a recent fad. Two hundred years ago, an average American consumed only 2 pounds of sugar a year. Even in 1970, the sugar consumption was 123 pounds per year, against the 152 pounds of sugar consumed in various forms today. Much of the sugar consumed by individuals is invisible to them because it is added to baked and processed foods.

Loss of Marine Lives

Marine life is in grave danger because of overfishing and pollution of waters. If humans continue fishing at the present level, there will be a global collapse of targeted fish species within a few decades.[301] Right now, oceans are losing their fecundity due to overfishing, pollution, and climate change. More than half of the oceans' surface, which is about four times the landmass covered by agriculture, is searched with ocean trawlers for seafood. Some countries, including Japan, Spain, China, South Korea, and the United States, provide substantial financial support to their fishing fleets to haul in as much marine life as possible. Several factors have combined to produce this state where the viability of seas as a source of food is in danger. Fish habitats have declined because many mangroves have been destroyed by fishing fleets. About a quarter of the marine life that lives on coral reefs will not survive due to warmer water and acidification caused by dissolved carbon dioxide—a process that has already started. Warming of the water in oceans due to climate change has an adverse effect on all marine life. Even a small increase in temperature will increase the acidity and change ocean currents. Some fishes will alter migratory patterns in search of cooler waters, which will disrupt their lives.

Besides the change in temperature and overfishing, marine life is also endangered by pollution. Since oceans are considered to be a limitless sink, all types of waste materials are dumped into them without consideration of their adverse effects. One of the major contaminants is plastic; there is already an estimated 250,000 tons in oceans. While some fish die by swallowing little plastic pieces, larger fish and marine mammals become trapped in nets and suffocate to death. Marine life, an important heritage of mankind, may mostly disappear in one or two generations if the present trend continues.

The Way Forward

Humanity's demand for animal-based foods is causing numerous environmental problems that threaten our welfare in many ways. The consumption of meats will continue to cause great damages to the planetary resources, adversely affecting our lifestyle, even our very existence. At present the livestock industry is one of the biggest industries in the U.S. and the world. Livestock contribute directly or indirectly to global warming, deforestation, species extinction, water pollution, air pollution, and desertification. In addition, it creates or exacerbates health problems in the human population, such as obesity and heart disease. The emissions of greenhouse gases from livestock and the production of feed make a large contribution to global warming. Animal agriculture necessitates the elimination of wildlife on a massive scale so that other animals may not compete with livestock for resources or endanger them, thereby decreasing the biodiversity that plays a crucial role in the ecosystem. Operations to eliminate competing wildlife are supported by the USDA and the Bureau of Land Management.

The livestock industry receives large financial support of $38 billion each year from the U.S. government, according to the USDA Agriculture Marketing Service. Without these subsidies, meat will be very expensive and out of reach of most American consumers. Most European countries also subsidize the production of meats by substantial amounts. Since the conversion of agricultural products to animal-based foods involves a large loss of energy and nutrients, a change in diet from meat-based to plant-based could add up to 49 percent to the global food supply without expanding croplands and will also significantly reduce carbon emissions and waste byproducts. Since water is used in all phases of livestock operations, reducing the consumption of animal-based foods would decrease the demand for water by at least 50 percent. This is an important consideration because the supply of water is limited in many parts of the world. Animal agriculture is the second-largest contributor to human-made greenhouse gas emissions, after the fossil fuel industry, and is a leading cause of deforestation, water and air pollution, and biodiversity loss. It would be impossible for a global population of 10 billion, which we are projected to reach by the middle of the century, to eat the amount of meat typical of the diets in North America and Europe because it would require too much land and water, leading to unacceptable levels of greenhouse gases and other pollutants in the environment.

Two main reasons for the precarious food situation, which leaves almost a billion people in the world hungry for some periods of the year, are increasing population and greater consumption of animal-based foods. Animal agriculture in its current form is simultaneously a driver of global environmental change and a victim of shifting environmental conditions. We need to seriously consider what we eat and the way it is produced. Strategies for increasing food production while alleviating environmental burden include reducing food losses, increasing agricultural productivity, producing animal-based foods only on marginal lands, and drastically reducing their consumption. Achieving a healthy and sustainable food system is an urgent matter due to increasing population and changing food preferences, and it requires collaborative efforts between governments, the private and public sectors, and individuals. Plant-based diets are better than animal-based foods both for health outcomes and the environment. In the end, what is good for the planet is good for our health: there is a correlation between eating healthy and eating climate-friendly. According to Walter Willett, professor of epidemiology and nutrition at the Harvard T.H. Chen School of Public Health: "We are presently on a path leading to a seriously degraded planet. If we care about the world our children and grandchildren will live in, we need to transform our diets and the way we produce out food. An immediate benefit will be improvements in our health and well-being."

SEVEN

· · · · · · · · · · · · · · · · · · · ·

WEALTH CONCENTRATION

A phenomenon of the last few decades is the concentration of wealth in an exceedingly small section of society in almost all countries of the world. The substantial economic growth in the U.S. from the end of the Second World War to the early 1970s was shared by all sections of society—the rich, middle class, and the rest. Although there were differences between the incomes of various sections of society, all of them prospered roughly by the same proportion and the percentage gain by each group was roughly the same. However, the era of shared prosperity ended in the mid-'70s with income growing much faster at the top of the income ladder than in the middle or at the bottom. The incomes of all lower and middle classes have stagnated since then, but the gains in incomes of rich persons have continued to increase, even accelerate. From 1970 onward, the share of income of the middle-class households decreased from 62 percent to 43 percent. During the same period, the proportion of income of upper-income households increased from 29 percent to 48 percent and the share of household income of the lowest-income households decreased from 10 percent to 9 percent. All these years, the income of the upper-income households has kept increasing at the expense of other sections of society.

These figures do not give a complete picture of the gains made by rich people in the United States. Wealth of families, the total value of a household's property and financial assets, is much more highly concentrated than income. In 2016, the top 1 percent of families owned 39 percent of the nation's wealth, the next 9 percent owned 38 percent, and the bottom 90 percent owned a mere 23 percent. The wealth of the richest persons in the country has steadily increased during the last few decades. While half the country's poorer families have less real wealth than they did twenty years ago, the super-rich (the top 1 percent) have doubled their wealth in a generation.[302] The top 1.2 million households had an aggregate net worth of $35 trillion as of June 2019, 32 percent of the total wealth in the country. At the same time, the poorest half of the population owned just 2 percent of the national wealth. In the United States, just three individuals—Jeff Bezos, Bill Gates, and Warren Buffett—own more wealth than the entire bottom half of the country.[303] The top 1 percent of American households control 35 percent of the nation's wealth, and the top 10 percent own a staggering 76 percent of the nation's wealth. The median household wealth in 2016 was $78,000, slightly lower in inflation-adjusted dollars than the wealth of $80,000 three decades ago. Over the same period, the average wealth of the top 1 percent household increased from $10.6 million to $26.4 million.[304]

The true wealth of the superrich cannot be determined because they hide their assets in offshore accounts and tax shelters such as Bermuda, the Cayman Islands, and the Virgin Islands, out of the purview of the government. Gabriel Zucman and colleagues calculate that offshore tax havens are enabling the world's richest 0.01 percent to evade 25 percent of the tax that they should be paying. Eighty-two percent of the wealth generated in 2017 went to the richest 1 percent of the global population while 3.7 billion people, one half of the world's population, had no increase in their wealth. At the global level, represented by China, Europe, and the United States, the top 10 percent own more than 70 percent of the total wealth while the bottom 50 percent own less than 2 percent of the global wealth.

Policies of governments are responsible for the extreme inequality in many countries. Tax rates for wealthy individuals and corporations in rich countries have been greatly reduced during the last few decades. Reducing the taxes on wealthy individuals often necessitates reducing expenditures on social services such as health care and education. Since most of the population depends on these facilities, a reduction in expenditures on them adversely affects their lives.

There are many pernicious effects of the lack of resources on the general population due to extreme wealth inequality—people often do not have access to quality health care and cannot afford nutritious food, which may even result in shorter life spans. Easy availability of low-cost, highly processed foods containing empty calories and little nutritional value has substantially increased the number of obese and overweight persons among the impoverished people in most developed countries. The number of obese or overweight people in Europe increased three-fold over the last twenty years.[305] In the U.S., the numbers of people who live in counties with high rates of poverty have much greater obesity rates than those in wealthy counties.[306] Important indicators of well-being such as life expectancy, infant mortality, and obesity can be linked to the level of prosperity in each population. Possession of wealth provides, in addition to luxuries, a passport to better health and longer life.

The wealth gap between the rich and the rest of the population is also expanding. The concentration of wealth in an exceedingly small segment of society gives them political power. Since the major corporations are owned or controlled by the superrich, expenditures by middle and lower classes on household items continues to increase their wealth. While the governments in many countries are reducing taxes on the wealthy and big corporations, public services are being reduced due to chronic underfunding.

The coronavirus pandemic of 2020–2021 greatly widened America's wealth gap and exacerbated income inequality. As the COVID-19 pandemic spread across the country, unemployment and financial crises decimated the fortunes of hundreds of millions of working people while greatly increasing the wealth of the superrich. While ordinary people suffered from the health and economic impacts of the pandemic, the fortunes of billionaires greatly increased. It is estimated that the combined wealth of billionaires in the U.S. increased from $2.9 trillion in March 2020 to $4.3 trillion as of February 2021. During roughly the same period, 26.1 million U.S. workers were hit by the economic downturn and were officially unemployed or employed with a significant reduction in salary.[307] Many businesses that depended on the inflow of customers were forced to close, leaving customers no choice but to go to big corporations. The pandemic worsened the already worrisome level of income inequality in the country. Its impact during the pandemic was not just linked to economic hardship—it was also deadly. States with more income equality were likely to report more COVID-19 cases and more fatalities.

Declining Upward Mobility

Upward mobility is an essential component of a progressive society, with a chance for each generation to be financially better off than the previous generation. To determine the trends in financial mobility in the U.S. during the last few decades, Raj Chetty and colleagues at Stanford University obtained access to 60 million tax records stretched over more than four decades. These anonymous records were analyzed to determine the change in financial status across generations. Their exhaustive work showed that children's prospects of earning more than their parents, which were as high as 90 percent for those born in 1940, gradually decreased to only 50 percent for those born in the 1980s.[308] The results of the analysis are shown in Figure 10. For persons born in the late '80s or thereafter, it is likely that their chance of doing better than their parents will be even lower.

The United States offers less social mobility than most other developed countries.[309] Canada and European nations, with their social safety nets, investment in public goods, and progressive tax policies, have greater social mobility than the U.S. In fact, Canada, Australia, Denmark, Sweden, and Finland are among the most socially mobile countries in the world. People who have been raised with the American dream—that life will get better if they work hard and use available opportunities—are going to be sorely disappointed. In fact, even maintaining the living standard of their parents will be exceedingly difficult for most people. This realization is driving many people in rich countries toward radical politicians who, instead of finding solutions, blame the present state on minority groups and recent immigrants.

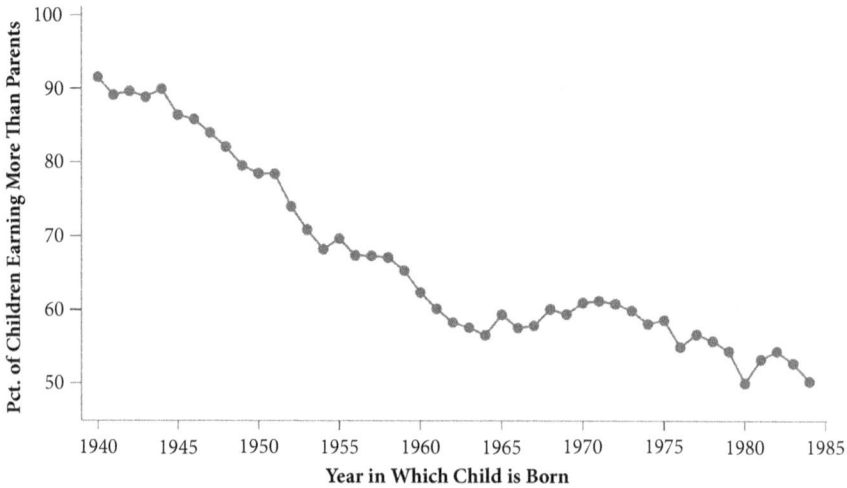

Figure 10 The Fading American Dream (Percent of Children Earning More than their Parents, by Year of Birth).

Source: Equality of Opportunity Project,[310] with permission.

Contributing Factors to Wealth Concentration

The concentration of wealth in a small section of society, as is happening in most countries of the world, results in lesser financial resources for the majority of the population. According to a Credit Suisse Report released in October 2020, the top 10 percent of households globally hold 43 percent of all personal wealth, while the bottom 50 percent own only 1 percent. Put another way, the richest 10 percent of adults in the world own 85 percent of the global household wealth while the bottom half collectively own barely 1 percent. Extreme inequality is now a global phenomenon. There are more billionaires than ever before and their wealth keeps growing with each passing year. The World Social Report 2020 published by the UN Department of Economic and Social Affairs (UN-DESA) shows that wealth inequality has increased in most developed and middle income countries. The concentration of wealth is also taking place in the most populous countries—China and India. In China, the richest 1 percent owned more than one-third of the total national household wealth while the poorest 25 percent owned less than 2 percent.[311] In India, the top 10 percent of the population holds 77 percent of the total national wealth.

The powerful global forces that are exacerbating inequality around the world are technological innovations, climate change, and urbanization. Rapid technological developents in the workplace benefit highly skilled workers, but those seeking manual or low-skilled jobs are left behind. Developments in technology are creating a digital divide between highly skilled workers and the rest. The climate crisis is having a negative impact on the quality of life, and vulnerable populations are bearing the brunt of environmental degradation and extreme-weather events. While climate change affects everyone, those involved in occupations related to agriculture are most seriously affected. The trend toward urbanization, occuring in most developing countries, affects the physical environment through the impacts of the number of people, their activities, and increased demand for resources.

International Trade

International trade deals have directly resulted in the decline in fortunes and powers of working people in many developed countries. The WTO requires member countries to substantially reduce tariffs and other obstacles to trade. When China joined the WTO in 2001, many multinational corporations set up manufacturing facilities in that country, moving them out of the United States. China has now become the manufacturing hub for many American companies because of its drastically lower labor costs. The net result of these trade agreements has been to transfer millions of jobs to China, Mexico, Vietnam, and other developing countries where labor is cheap and environmental regulations are either lax or nonexistent. Since 2000, the United States has lost 5 million manufacturing jobs to China and another million jobs to Mexico.[312] American manufacturing workers, who generally do not hold a college degree, make about $20 per hour and solidly belong to the middle class because their earnings are much more than the minimum wage in all fifty states. Loss of manufacturing jobs caused by the free trade agreements has been disastrous for these workers and has weakened the financial standing of a large section of society.

In addition to siphoning well-paying jobs to other countries, international trade agreements have also diminished the power of workers in developed countries. With the threat of their companies relocating to other countries hanging over their heads, employees are not able to lobby for higher wages or better working conditions and are forced to accept whatever the management decides to give to them, however small. This has stagnated the wages of workers in developed countries for a few decades and their median household income has not even kept pace with inflation. The benefiaries of multinational trade deals are large corporations and the wealthy people who own them. Due to the transfer of manufacturing hubs to other countries, America imports about half a trillion dollars worth of goods more than it exports, thereby causing a huge trade deficit. This imbalance is bad for the economy because it reduces the national GDP and resources of the government, thus hurting the middle class.[313] A continous trade deficit amounts to a stimulus to the countries that export goods to the United States. During the last two decades, the American economy has provided a stimulus package of more than $10 trillion to the rest of the world.[314] Millions of jobs have been sucked out of the U.S. economy by trade deals that let corporations replace American workers with low-paid, often mistreated, workers in other parts of the world.

Decline of Labor Unions

From their inception, labor unions fought with the corporations and even the government to protect the common interests of workers. During the decades after the Second World War, the struggles of organized unions achieved numerous benefits for workers such as the forty-hour work week, two-day weekend, lunch breaks, overtime pay, health benefits, and child labor laws. Many such achievements, which have now become accepted in all sectors, were strongly opposed by businesses. Between 1946 and 1967, the productivity of American industries grew 104 percent. A strong labor movement and tacit support from the government ensured that the wages of workers increased by roughly the same amount. The disconnect between productivity and wages began in 1973 and has increased with each passing year. During the forty-year period from 1973 to 2014, productivity increased 72.2 percent, while a typical worker's hourly compensation increased only by 9.2 percent.[315] This divergence between productivity and wages means that the profits enriched the high-level managements and further increased their wealth.

While corporations have always wanted to weaken the unions, they received support, or at least acquiescence, from the government when business interests gained influence and helped pass laws against unions. Despite the objections of President Harry S. Truman, the Taft-Hartley Act became a law in 1947, reducing disruptions caused by strikes and imposing restrictions on the activities and power of unions. Labor leaders called it the "slave labor bill." The passage of this law whittled down the power of unions. However, the seminal moment in the decline of unions was the firing of 12,000 striking members of the Professional Air Traffic Controllers Organization in 1981 by President Reagan. He also appointed ultraconservatives to the National Labor Relations Board, an independent federal agency that protects the rights of private sector employees to join together—with or without a union. In most conflicts, the new appointees sided with the management and against the demands of workers. The Reagan administration also created a long list of prohibitions on actions by unions and their organizers. The administrations after Reagan did not take any actions to support unions. Lack of governmental support and continuous attacks from management kept decreasing the strength and power of the unions.

With the decline in the power of unions, corporations began attacking labor unions in a number of ways. Strike breakers were hired in the event of strikes and workers were kept out of the factories after the strike. With their deep pockets, corporations also funded major lobbying efforts in Congress to ensure that the government would not act in a way that benefited labor. The diminishing power of unions is evidenced by the fact that the number of strikes decreased from 371 in 1970 to only eleven in 2010.[316] Businesses have also succeeded in reducing the financial resources of union organizers by helping to advance right-to-work (RTW) laws that have now been passed by twenty-eight states in the country. Under these laws, employees in unionized workplaces may not be compelled to join a union nor be forced to pay union dues while generally receiving the same benefit as union members. RTW laws greatly reduced the membership and finances of unions. Detailed analysis has shown that the wages of full-time workers are lower by $1,558 per year in RTW states than in non-RTW states. Workers in RTW states are also less likely to have employer-sponsored health insurance and pension coverage.[317] The latest RTW law was signed into law by Wisconsin Governor Scott Walker.

The United States Supreme Court has recently passed rulings that further cripple the power of workers and their unions. According to a judgment on May 2018, employers can now force workers to give up their right to sue collectively and seek one-on-one arbitration. Employers prefer such arbitration because lawyers are reluctant to contest a case in which the judgment or settlement will be small. In June 2018, the Supreme Court ruled that non-union workers cannot be forced to pay fees to public sector unions. This decision may make it difficult for public sector unions to survive because they will not be able to collect dues from all workers. This is yet another obstacle in the path of working men and women to collectively assert their rights with the power of a unified voice. This decision will reduce the financial base of unions by millions of dollars. The number of unionized workers has been decreasing continuously during the last few decades. At its peak in the 1950s, about 35 percent workers in the private sector were unionized. Although at present 11.3 percent of workers are unionized, this number includes police, teachers, and workers in the public sector. The number of unionized workers in the private sector has decreased to 6.6 percent.[318] A decline in the membership of unions reduces their influence at the bargaining table. As the power of unions decreases, corporations feel free to pay minimal wages to the workers and hoard the bulk of profit, thus contributing to a greater concentration of wealth at the top.

According to research by Pew, roughly eight out of ten Americans support the right of workers to form unions to counter the excesses of corporations.[319] An analysis by economists at the IMF shows that a decline in unionization lowers the wages of all workers and increases the income of senior executives.[320] Thus, the impact of declining unionization is felt across the entire income spectrum because it reduces the welfare of workers and transfers wealth to corporate managers. Strong unions are needed to sustain a solid working class so that they can counter the political power and influence of corporate America. Although the decline of unions is often blamed on international trade and globalization, unions remain strong, certainly much stronger than in the U.S., in many other industrialized countries such as Germany, France, and Canada.

Automation

Machines began replacing humans from the beginning of industrialization. Despite the fears of Luddites, most jobs lost to automation in those days were replaced by other types of jobs. The introduction of robots and artificial intelligence (AI) affects jobs in a completely different way. With access to large databases and logic to make intelligent decisions, the number of jobs that can be replaced by automation and smart machines is very large. Bank tellers, telemarketers, and customer service representatives have already been replaced in many establishments. In repetitive tasks, robots have an edge over humans because of their greater accuracy and endurance, thus leading to greater productivity. While early stages of automation were focused on manual repetitive tasks, the capabilties of these machines are increasing at such a rapid pace that they may replace humans even in complex situations.

A substantial amount of trading on Wall Street is done by automated algorithms. They have advantages over humans because of their high speed and instant access to a large database. Self-driving cars have already been tested in real-life situations, and the development of trucks that follow a specific route to transport goods over long distances is on the horizon. Driverless vehicles will wipe out millions of solid middle-class jobs and destroy numerous businesses. Even high-end jobs may not be immune from automation. Robots already assist doctors in surgery and a few hospitals are testing robots to assist or replace doctors for routine care. Online teaching is becoming more and more popular since it reduces the cost of maintaining a physical facility. More than 60 percent of the 173 occupations projected to decline through 2021 are well-paying jobs, according to an anlysis by CareerBuilder's Economic Modeling Specialists International.[321] Economists Frey and Osborne of Oxford University have calculated that 47 percent of the total U.S. employment, involving more than 50 million persons, is at risk of automation within the next decade or two.[322]

Using robots in the workplace only requires an initial investment and occasional upgrades. Since robots can work faster and with greater efficiency and accuracy than humans, their use increases productivity without increasing recurring expenses. For these reasons, corporations and large employers support the development of robots. By reducing expenditure on workers, robots greatly increase the profit of businesses while taking jobs away from humans. Development of robots, which is being done at an ever-increasing pace, will cause the loss of a much greater number of jobs than the jobs lost in trade deals. AI and robotics in the workplace could cause mass unemployment and dislocate entire economies. Automation is one of the reasons for the accumulation of wealth by corporations and their senior executives. However, artificial intelligence is still in its infancy. Full-fledged development of AI and the installation of robots may take a few decades. At that time, the role of robots will advance from routine physical tasks to non-routine cognitive tasks such as those of teachers, doctors, and executives. When that happens, the beneficiaries will be the big capitalists who own the robots and employ them in their business. The rest of the population who trade work for money will be the losers because there will not be many things left for them to do.

Elon Musk, founder and CEO of Tesla, is among the many leaders who are extremely concerned with the development of artificial intelligence. He describes AI as "the greatest risk we face as a civilization."[323] While the government regulates the safety of cars, airplanes, food, and drugs, the AI industry, which "is a fundamental risk to the human civilization" is not regulated. He calls for swift and decisive actions to regulate developments in the industry in a proactive—rather than reactive—manner to control the growth and deployment of such devices.

Power of the Superrich over the Political Process

The rich have pulled away from the middle class because of their political power. Perhaps the most corrosive aspect of the concentration of wealth in a small section of society is that they can directly or indirectly control the political system so that their wealth and power keep growing at the expense of other segments of society. High-priced lobbyists and unlimited campaign contributions give them an outsized voice in the political process. Extreme inequality essentially disenfranchises the majority of the population because the power of voting is insignificant compared to the power of money on legislators. The superrich use their clout and access to politicians to control the political process so that their advantages, tax breaks, and control of public assets not only continue and but also keep increasing.

Martin Gilens of Princeton University and Benjamin Page of Northwestern University analyzed twenty years of data from nearly 2,000 public opinion surveys and compared it to policies that eventually became laws. They found that organized groups representing business interests essentially determine the policies of the U.S. government while the opinion of the bottom 90 percent of income-earners has essentially no impact on governmental policies.[324] There is a synergistic relation between extreme concentration of wealth and political power; the superrich influence the political process to increase their wealth, and more wealth further increases their influence on the government. Observations of this type show that the government mainly represents the moneyed interests and not ordinary citizens. A consequence of the superrich population's control of the political process is that rest of the people feel disconnected from the governmental actions and do not get involved by contacting their representatives and even abstain from voting, thus further reducing their influence in shaping the direction of national policies.

The growing disparity between the rich and the rest of the population touches every aspect of daily lives. The difference between the superrich and others is not only that they live in palatial homes, reside in exclusive communities, and travel by private planes, but that the differences permeate all walks of life with far-reaching effects. The quality of public schools depends on the average wealth of residents in the neighborhood, thus adversely affecting the prospects of students from poorer neighborhoods. Residents of less affluent neighborhoods are also shortchanged in health care, social services, safety, and other needs. These differences tend to perpetuate the advantages of the wealthy and lock families of lesser means into their inferior status. The concentration of wealth in a small percentage of households has damaged the American dream more than the slowdown in economic growth.

Some of the accumulated wealth of the richest people is used in financial gimmicks through mergers and acquisitions that increase their wealth without any material gain. Despite claims by some pundits who support trickle-down economics, a large fraction of the accumulated wealth does not create jobs and is not circulated in the country but is parked in low-tax or no-tax countries like the Cayman Islands, Ireland, and Switzerland—out of the purview of American authorities. Hiding wealth from national authorities is done by wealthy people in many countries. James Henry, former chief economist at McKinsey, believes that $21 trillion has been parked by the superrich in various countries in secret accounts in tax havens. This amount is greater than the GDP of the United States and Japan put together.[325] While the profits of their operations are stored in offshore accounts, the losses are shown in American branches, thus reducing or eliminating the taxes owed to the country. Using various kinds of financial manipulations, the online retail giant Amazon did not pay a dime of federal income tax on $5.6 billion in profits in the United States in 2017.[326] Storage of such a large amount of wealth in nonproductive vaults makes a lie of the basic premise of "supply-side economics" or "trickle-down economics" that claims that a concentration of wealth is good for the country because "a rising tide lifts all boats."

Remediation of the Present Situation

The first step toward any remediation of the present situation is the recognition that hundreds of millions of people are not able to realize their full potential. Financial constraints force them to work long hours in mundane jobs that do not fully use their abilities or potential. The waste of talent at this scale is not only a loss for the individuals, but also for the country at large. We have reached this stage due to concerted efforts by the superrich and their corporations to bend the rules in their favor. This powerful group now has a hold on Congress through direct or indirect financial contributions. The lobbyists who work as their agents now have the power to write or modify laws, ratified by elected officials that benefit their paymasters.[327] Public opposition to the firmly entrenched financial bigwigs will not have a significant effect. The only way to change the system is to elect representatives who truly believe in the ascendancy of the middle class and, perhaps more importantly, are willing to work for it. We need leaders in the mold of FDR and Lyndon Johnson who consider the welfare of the middle class to be a high priority item on their agenda. The struggle to reverse this course and save the middle class requires focus on specific issues that will tangibly support a large segment of the population, some of which are detailed below.

(1) Higher Taxes on the Wealthy

We have lived with the fallacy of lower taxes and smaller government—the pillars of supply-side economics—for a few decades now. The result is an accumulation of wealth by the richest Americans and lower quality of social services such as health care, education, and public transportation. The progressive tax code of the 1960s and '70s has been modified a few times, starting with the Reagan administration, to decrease the tax liability of the richest Americans. In addition, the tax code is riddled with loopholes so that the superrich often pay taxes at a lower rate than middle-class families. The demand for lower taxation resonates with the impoverished middle class even though the taxes paid by Americans are among the lowest of all developed countries.[328] At the same time, our infrastructure is deteriorating—numerous roads, bridges, and tunnels are in bad shape. Some localities do not have access to vital resources such as clean water, whereas others are behind on technology and innovation. For example, the quality of internet service in much of the country is poor compared to what is available in many European and Asian countries. In general, public transportation is either very expensive or nonexistent. To ameliorate the situation and release funds for projects and services that will help everyone, it is important to raise taxes on big corporations and wealthy individuals. Warren Buffett, one of the richest men in the country, has advocated a minimum tax of 30 percent on rich people.[329]

Another important step in advancing societal equality is taxing income from investments at the same rate as taxes on wages. The lower tax rate on investment income has given a large windfall to wealthy Americans, estimated to be about $1.3 trillion during the last ten years, and is a key driver of economic inequality. Due to this clause in the tax code, some superrich Americans pay income tax at a lower rate than many professionals. When the wealthiest Americans reduce their taxation by loopholes, the rest of the population must pick up the slack. The revision of the tax code in 2018 left the tax rate on dividends and capital gains unchanged. The new rules allow a 20 percent deduction for pass-through income—income which is generated through a business, but "passed through" to owners to be taxed under individual income tax instead of the corporate tax. Although "pass-through" is a term often used to refer to small businesses, a Treasury Department Analysis found that 69 percent of the pass-through income actually goes to the top 1 percent of households.[330] Carried interest is another provision of the tax code that is used by hedge fund owners to reduce their tax liability. President Trump criticized it by saying "Hedge fund guys are getting away with murder." However, the Tax Cuts and Jobs Act of 2018 left this provision intact.

(2) Universal Health Care

Almost all rich countries provide universal health care, although the actual implementation is somewhat different. Americans spend about 17.1 percent of the GDP on health care, while still leaving about 30 million persons uninsured. Even with full healthcare coverage to all citizens, the proportion of GDP spent by other OECD countries is much less. While Sweden, Germany, Denmark, and New Zealand spend about 11 percent of their GDP on health care, residents of the U.K. spend only 8.8 percent of GDP. Annual healthcare spending in the U.S. is more than twice the OECD median. Private spending on health care for things like copayment and prescription drugs is the highest in the U.S. Still, the United States lags behind other countries in important indicators of national health such as life expectancy, infant mortality, and obesity.[331]

There are two main reasons for the high cost of health care in the United States. Unlike socialized health care in all rich countries, private insurers cost a lot of money without adding any value. Switching to single-payer health care will save billions of dollars in healthcare expenses. Recent scholarship shows that a single-payer national healthcare program would provide comprehensive coverage to everyone in the country without copayments or deductibles. All medically necessary services would be covered, including prescription drugs.[332] In addition to providing health care to all citizens, a single-payer system would also ensure that no one becomes destitute because of healthcare expenses. The barrier to universal health care is not economic but political.

The cost of health care in the country has increased from $74.6 billion in 1970 to $3.34 trillion in 2016, far outpacing the increase in the average worker's earnings or overall inflation. Employers are now passing on a greater proportion of the costs to employees in the form of higher premiums and greater deductibles. In the United States, the per capita cost of health care is $11,582 per year, while the average amount spent on health per person in comparable countries is only $5,697.[333] During the last twenty years, American healthcare costs have increased at a much faster rate than other economic indices such as GDP per capita, median household income, or inflation. During the 1960s and 1970s, employers picked up most of the cost of their employees' medical expenses. However, with increasing costs of medical care, employers are now passing on an increasingly greater share of costs to employees. In 1973, 66.7 percent of recent college graduates and 23.5 percent of high school graduates had health insurance covered by their employers. These proportions have gradually decreased to 30.9 percent and 6.6 percent, respectively.[334]

(3) Estate Tax

The lower limit of the estate tax—the tax on the transfer of the estate of deceased person—was increased by the Trump administration from a base of $5 million set in 2011 to $10 million for tax years 2018 through 2025. The exemption is indexed to inflation, so an individual can shelter $11.2 million in assets from estate tax and a couple could exclude $22.4 million.[335] Even before the Trump administration's increase of the estate tax's lower limit, only 5,000 estates out of approximately 2.7 million deaths each year in the United States paid any estate tax. The estate tax of 40 percent is levied on estates above this limit, making the amount relatively small on a proportional basis. This increase in the lower limit of the estate tax will cause a loss of billions of dollars to the treasury. Hillary Clinton, while running for president, proposed to lower the threshold to $7 million for couples and to impose a tax of 45 percent above that threshold. Proponents of further decreasing the estate tax argue that it forces farmers to sell their farms. However, this claim is not backed by data: only fifty farms or closely held family businesses in the U.S. would have paid estate taxes before the increase in the exemption limit in 2018.[336] An effective increase in estate tax will also help in decreasing the grotesque wealth inequality in the country.

(4) Banking Changes

The Glass-Steagall Act was passed after the Great Depression in 1933, separating investment banking from retail banking. The Act was repealed in 1999 by President Clinton with active support from Republicans. It triggered a series of international mergers that created companies that were so vital to the global financial system they were considered to be "too big to fail." It is widely believed that the financial crisis of 2008 was caused by speculative banking allowed by the repeal of the Glass-Steagall Act. It is important to split retail and investment banking so that banks do not indulge in speculative investing without the consent of investors. There is no reason to commingle banks that deal with savings, checking, and mortgages with those that deal with hedge funds, derivatives, and other high-risk items. Failure to separate them may lead to another financial crisis that will shake up the economy.

(5) Public Financing of Elections

One of the reasons for the undue influence of wealthy people on politicians is that contesting an election is a very expensive proposition these days. The Supreme Court's decision in Citizens United vs. FEC in 2010 opened the floodgate of money into American elections. After this, in Arizona Free Enterprise Club v. Bennett, the Supreme Court ruled that additional public grants made available to a publicly funded candidate violated the rights of those who oppose publicly funded candidates. Almost every election now breaks the financial record set by the previous election.[337]

Due to the high expense of contesting elections, politicians must keep raising money all the time. This process makes them beholden to individuals or organizations that provide them with the necessary finances. Passing laws that help their benefactors ensures a continuous flow of funds. The army of lobbyists in Washington D.C. influences, even dictates, the political process. The best way to eliminate this corrupt practice is to provide public financing of all elections. In addition to freeing the politicians from undue influence of donors, publicly funded campaigns will also foster diversity in the electoral process and encourage voter-centered campaigns.

(6) Shorter Workweek

The forty-hour workweek was introduced by Ford Motor Company in 1914 because Henry Ford believed that working for more hours negatively affects the productivity of workers. It became a standard for American businesses after a struggle between unions and owners of factories that lasted for almost twenty years. The duration of the workweek must be reevaluated because of changes in modern job markets. Automation and the applications of artificial intelligence have decreased the need for human labor. This development has already reduced the need for bank tellers, travel agents, and retail marketing personnel. It is projected that 47 percent of jobs in the United States are at risk of elimination in the near future. In addition, a worker today has to be much more knowledgeable than in the past and must also continuously upgrade his or her skills to stay abreast with new developments.

The national wealth has been steadily increasing for the last few decades but the benefits from the gain in productivity have simply been absorbed by corporations and their senior executives. A reduction in working hours to thirty, and finally to twenty, while keeping the compensations the same, will help millions of workers, and also reduce the pace of growing inequality. Scientific studies have shown that it is difficult to remain creative after more than six hours of work in a day. Reduced work hours without reducing wages will lead to lower stress, reduced societal polarization, connected communities, and healthier people because workers will have time for family, community involvement, and leisure. In the free time released by this change, workers may be able to learn new skills and trades that will benefit them in future employment.

(7) Raising Minimum Wage

The federal minimum wage has remained unchanged at $7.25 per hour since 2009, although many states have started increasing the minimum wage in small steps. The minimum wage is not a survival wage, and most families use welfare (TANF), food stamps (SNAP), and earned income tax credits to make a living. Most employees of highly profitable companies like Walmart, McDonald's, and Amazon use some of these programs. These safety-net programs cost federal and state taxpayers more than $150 billion per year.[338] Thus, the operation and profit of low-wage employers is subsidized by the general public. It is important that the federal minimum wage be increased by a substantial amount so that the minimum wage is a living wage without subsidy from public funds. Raising the minimum wage will put pressure on other low-paying jobs to increase their employees' salaries. In addition, this change will also allow the Social Security trust fund to remain solvent for a longer time. With an increasing population of elderly people, another step that should be taken is to remove the cap on Social Security withholdings. President Biden is considering increasing the minimum wage.

(8) Clemency for Student Loans

A major problem that is preventing a large section of society from fully participating in the economy is the burden of student loans. A total of 45 million borrowers across the country now owe $1.71 trillion to the lenders.[339] Over the next decade, agencies of the federal government will make $110 billion in profit on student loans—paid by people struggling to join the middle class. It is important to immediately reduce the interest rate on student loans and forgive the loans after a person has faithfully paid the installments for ten years. Public colleges and universities should be made free for all residents of the state. Student loans get a favorable treatment in many developed countries, including Great Britain, Denmark, Norway, Sweden, and Australia, where they are tied to the salaries of the borrowers and balances are forgiven after a certain number of years. The burden of student loans adversely affects the economy because a substantial amount of the earnings of young people is used for repaying student loans. It delays the entry of millions of people into the middle class by tying their earnings to loan repayment. The enormous amount owed by a large segment of the population is a drag on the economic growth of the country and leads to heightened financial insecurity among citizens. The cost of higher education and difficulties in repayment even discourages many persons from pursuing higher education.

(9) Basic Income

The concept of basic income—supported by industry leaders such as Bill Gates, Elon Musk, Mark Zuckerberg, and Sir Richard Branson—involves providing a limited income to all citizens regardless of professional status or intent to find work. Basic income will allow people to develop new ideas and explore business prospects without putting their livelihoods at risk. People receiving basic income will be free to take employment or pursue any profession that interests them. It will eliminate the bureaucracies and expenses of social welfare programs such as welfare, food stamps, and unemployment insurance. By providing a certain amount to all citizens, it will eliminate extreme poverty. The cost of giving $10,000 per year to all citizens is estimated to be $3.2 trillion. If we exclude retirees and people making more than $100,000 per year from the recipients of basic income, the required amount decreases to $1.5 trillion. Such a program is being considered by Finland, Switzerland, Kenya, India, Netherlands, Canada, and some other countries.[340,341] Universal basic income appears to be the best answer to unemployment caused by automation as artificial intelligence decreases the availability of jobs.

The Way Forward

Improving the financial status of the great majority of people will require collective actions by the body politic of the country, which may involve a conflict with the entrenched interests of the wealthy and powerful. They will resist changes in the present system with all their power, including the media that they control. The superrich and the financial bigwigs already control some media outlets and are trying to control more.[342] It is important to know that economic issues are central to our well-being. An educated citizenry can prioritize complex issues and demands. There are many issues, other than the economic ones, that are important and rightfully championed by a growing number of passionate individuals. These include, among others, LGBTQ rights, freedom of choice, immigration, women's rights, and foreign policy. At the time of elections, it is common to deflect the attention of the electorate by raising these issues as a diversionary tactic and make them supremely important if it helps elect a candidate who supports maintaining the status quo in economic matters. Intelligent people can have differences of opinion on these topics. However, when the financial well-being of a large section of society is at stake, economic issues take precedence over others. It is important to educate the electorate to prioritize these issues and leave debatable issues for later developments. These are going to be hard-fought battles that may not yield results in a short time.

EIGHT

·····················

OVERPOPULATION: THE ELEPHANT IN THE ROOM

Most people do not want to discuss overpopulation because they consider procreation to be a fundamental right on par with other basic human rights. A discussion of regional variations in the rates of increase in population runs the risk of offending some groups. However, many problems discussed in previous chapters, and many of the challenges faced by humanity, are either caused or exacerbated by overpopulation. The population of the world reached a new high of 7.8 billion in December 2020 and is increasing almost everywhere. The U.S. population, currently 326 million, is projected to reach 398 million by 2050.[343] The rate of population increase is high—from 4 percent per year to 1.2 percent per year—in some of the poorest countries.[344] For a variety of reasons, the growth in population has not been the same in various parts of the world and for different groups within the same demographics.

Current projections are that the human population will be 9.6 billion by the year 2050, and further increase to 11 billion by the end of the century.[345] Most people desire a future with a better standard of living for their children, which inevitably means a greater consumption of planetary resources. However, even the current population of 7.8 billion and our present lifestyle are causing numerous environmental damages that are becoming more and more significant with the passage of time. Primarily due to the pressure of population and a lifestyle that does not consider sustainability to be a priority, the global ecosystem has come to a precarious state. Signs of the overexploitation of the ecological resources are everywhere. Climate change, which will eventually threaten the welfare and survival of humanity, is progressing at a steady pace. Extremes of weather have become common, sea levels are rising, mean global temperature sets a record almost every year, and the productivity of seas to provide us with marine life is seriously compromised. Directly or indirectly, most of such developments are related to the overpopulation of humanity.

Overpopulation occurs when the land cannot support and provide for the needs of the population of that region without serious environmental degradation. Continuation of this process will cause shortages of food, clean water, and amenities that are necessary for life. As the population increases, additional land must be brought into cultivation by razing forests, which will increase the load of greenhouse gases in the atmosphere, thereby making a greater contribution to climate change. Overpopulation greatly increases the need for fresh water for direct consumption and daily use, and a much greater amount for growing food. According to various projections, 50 percent of the global population may live in water-stressed regions by 2030, which is directly attributed to population growth.[346] An increase in population will inevitably create pressures leading to deforestation, decreased biodiversity, and spikes in pollution and emissions that will lead to severe ecological disruptions, perhaps even collapse. In an extreme case, the viability of human life on the Earth may be endangered.

The present population of the world is already making it difficult to provide the necessities of life to people in many places. Planetary resources have been stretched to the limit, still leaving a significant proportion of the population without the requirements of food and water, let alone a comfortable life. The impending climate change has begun to affect lives with extreme-weather events, higher temperatures, forest fires, floods, and droughts. Global warming is expected to have a far-reaching, long-lasting, and, in many cases, devastating impact on people everywhere. Most of the environmental problems, such as climate change, loss of species of both plants and animals, and shortages of planetary resources are either caused or exacerbated by population growth during the last few decades. According to the organization Population Connection,[347] population growth since 1950 is behind the clearing of 80 percent of rainforests, the loss of tens of thousands of plant and wildlife species, an increase in the emission of greenhouse gases by about 400 percent, and the development or commercialization of as much as half the Earth's surface land. The group fears that in the coming decades half of the world's population will be exposed to water stress and water-scarce conditions.

While population numbers in most developed countries are leveling off or diminishing, high levels of consumption by them makes a huge drain on planetary resources. Americans, for instance, who represent only 4 percent of the world's population, consume 25 percent of all resources. Industrialized countries contribute far more to climate change, ozone depletion, and loss of planetary resources than developing countries. The consumption of the world's wealthiest 10 percent produces up to 50 percent of the planet's consumption-based CO_2 emissions, while the poorest half of humanity contributes only 10 percent. Developing regions in Africa, Asia, and Latin America bear the brunt of climate and ecological catastrophes despite having contributed the least to them. The problem of extreme inequality, the excessive consumption of the world's ultrarich, and an economic system that prioritizes profit over social and ecological well-being are to be blamed for many of the current problems.[348] Many residents of developing countries emulate the over-consumption-heavy lifestyle that they see on television or the internet. A paradox of lower fertility and reduced population growth is that as education and affluence improve, the burden of natural resources increases per person. As people grow richer, each person consumes more natural resources and energy, typically from carbon-based fuels such as coal, oil, and gas. This lifestyle change with increasing wealth and education includes higher protein foods such as meats and dairy, more consumer goods, bigger homes, more vehicles, and more air travel.

The table below shows the changes in the population of different regions of the world from 1950 to 2018.

1950 (Total Population 2.64 billion)[349]	
Asia	55.4%
Europe	21.7%
Africa	9%
North America	6.8%
Latin America and Caribbean	6.7%
Oceania	0.4%

2018 (Total Population 7.63 billion)[350]	
Asia	59.5%
Africa	16.9%
Europe	9.7%
Latin America and Caribbean	8.6%
North America	4.8%
Oceania	0.5%

These numbers show that about 60 percent of the world's population now lives in Asia. The percentage of persons living in North America and Europe decreased, not because their absolute numbers declined during this period, but because the populations of Asia and Africa increased much more rapidly. Although the U.S. does not collect data about religions, Pew Research Center determined from a large-scale study that there are about 70.6 percent Christians of all denominations, 5.9 percent Jewish, 1.9 percent Muslims, and 22.8 percent unaffiliated in the U.S.[351] The projections suggest that the Muslim population in the U.S. will grow much faster than the country's Jewish population and they will replace Jews as the nation's second-largest religious group after Christians by the year 2040. The U.S. Muslim population is projected to reach 8.1 million, or 2.1 percent of the nation's total population, by 2050. Religious conversions did not have a large impact on the size of the U.S. Muslim population, largely because about as many Americans convert to Islam as leave the faith.

There are wide differences in the population growth of various religious groups in the world, which must be taken into consideration in a discussion of sustainability. Considering the period up to 2050 when the population in the world is expected to reach 9.8 billion, the proportion of Christians will remain roughly the same at 31.4 percent, but the population of Muslims will increase from 23.2 percent of the world's population to 29.7. Followers of the Hindu religion will almost retain their proportion at about 15 percent, and the unaffiliated, Buddhists and minority religious groups will decline by small amounts. On a worldwide basis, the rate of increase of the Muslim population is twice as large as that of Christians or Hindus because of greater fertility and a youthful population. For the period 2015–2060, while the overall growth in human population is expected to be 32 percent, the growth of the Muslim population will be 70 percent, larger than that of Christians (34 percent), Hindus (27 percent), and Jews (15 percent). The total number of Muslims in the world is projected to increase from 1.8 billion in 2015 to 3 billion in 2060. Muslim women have more children, an average of 2.9, significantly above the next highest group of Christians with 2.6 and the average for all non-Muslims of 2.2.[352] In all major regions where there is a sizable Muslim population, the fertility of Muslims exceeds that of other groups. In India, the number of Muslims is growing at a faster rate than the country's majority Hindu population and is projected to increase from 14.9 percent of India's population in 2015 to 19.4 percent in 2060. And while there were similar numbers of Muslims and Christians in Nigeria in 2015, the Muslim population is expected to grow to a solid majority of 60.5 percent in 2060. The total number of Muslims in the world is projected to increase from 1.8 billion in 2015 to 3 billion in 2060. In all major regions wherever there is a sizable Muslim population, Muslim fertility exceeds that of non-Muslims.

Considerations of religious distribution become important because, in today's interconnected world, no people live in isolation and an unrestricted growth of any group affects the resources available to all people, even in distant places. Supporting the present population in the world is already causing great harm to the ecosystem, which will greatly increase with an even greater population. We are running out of fresh water—a resource that is essential for our survival. Forests that provide numerous services are being razed to meet some immediate needs. The loss of habitat due to encroachment by humans is causing a mass extinction of numerous species, thereby endangering the complex web of life on the planet. Human activities and demands are causing an immense loss of marine life in seas and oceans. The use of fossil fuels is changing global climate, which may lead to, besides many other dangerous things, a rise in sea levels that would affect the lives of millions of people. Many other things that adversely affect the welfare of humanity can be added to this list. Increasing population will put additional pressure on the planetary resources. In extreme cases, it may even endanger the lives of a large number of people.

The primary causes of an increase in population of any group or region are fertility of women and the proportion of young women in the population who have their prime childbearing years ahead. The global average fertility is 2.5 children per woman but there are wide variations in different parts of the world. Fertility is highest in Africa at 4.7 children per woman, and Europe has the lowest fertility of 1.6 children per woman. Asia, Latin America, and the Caribbean have a total fertility of 2.2 children per woman. The fertility of women in East Asian countries is only 1.6 children per woman.[353] As of 2015, the Muslim fertility rate for all forty-nine Muslim countries was 2.9, well above the global average. The primary reasons for the greater fertility of women in Muslim countries include ignorance and lack of access to contraceptives. Although some progress has been made in the availability of contraceptives and the attitude of women toward birth control, the progress has been very slow.[354] The demographic transition from high to low fertility is particularly small in rural areas.[355] In democratic countries where Muslims are in a minority, they have an incentive to change the demographic with greater numbers. In May 2017, Prime Minister Recep Tayyip Erdogan of Turkey exhorted Muslims to "make not three but five children because you are the future of Europe."[356] The population of Muslims in India increased from 9.6 percent of the total population in 1950 to 14.2 percent in 2018, indicating a much greater rate of growth than the rest of the population. It is estimated that by 2050, India's Muslim population would grow 76 percent while the population of Hindus will increase by 33 percent.[357]

The scarcity brought about by overpopulation can cause political unrest and has the potential to trigger an increase in violence. We are already seeing wars fought over water, land, and energy resources in the Middle East and some other regions, and the turmoil is likely to increase as the global population grows even larger. So far, only a tiny section of the human population has shown a willingness to live within the ecological limits of the planet and reject chasing economic growth in the interest of stabilizing the climate or prevent the further destruction of the ecosphere.

The Way Forward

The most important way of dealing with the problem of overpopulation is to educate and empower women, and to provide them with opportunities to participate in the economy. This will involve imparting family planning education to young people through schools. It is important that young people, particularly girls, understand how their lives will be improved if they do not wish to spend a large part of it taking care of children. Educated and working women are more likely to support limiting family size by using birth control. Young men should also be educated on how not being tied to young children will free them to do other productive things. In general, the prosperity of a family should be decoupled from the number of children. However, this requires that opportunities should be provided where young people can channel their energies, including career opportunities. At the same time, sensitivity to the global problems created by overpopulation must be stressed. Governments can also provide incentives for limiting the size of families and disincentives for having more children. Studies show that women with access to reproductive health services find it easier to break out of poverty and those who work are more likely to use birth control.

APPENDIX

FOSSIL FUELS

Fossil fuels and their derivatives have essentially ushered in the modern age. While the direct combustion of coal is still used in some places as a source of energy, many power plants everywhere produce electricity from this basic raw material. In addition to a source of energy, fossil fuels are used to make numerous chemicals, including plastics and agricultural chemicals. It is nearly impossible to live these days without the use of fossil fuels in some form or the other.

There are three types of fossil fuels—coal, natural gas, and petroleum—that are used in different ways. All of them produce energy and emit green-house gases but have different uses and effects on the environment. Although coal has been recognized as a source of condensed energy since ancient times, its large-scale use in the early 20th century was instrumental in ushering in the industrial revolution. Its primary use today is in power plants that produce electricity, but some coal is used as a direct source of energy for household use, particularly in developing countries. Natural gas—usually found close to crude oil—is delivered through a network of pipes and supply lines or directly delivered to consumers in cylinders. In addition to being a convenient source of heating for domestic and industrial purposes, natural gas is also used for making fertilizers. Crude oil or petroleum has the greatest impact on our lives. Various components of crude oil, separated by fractional distillation, provide numerous substances such as heating oil, aviation fluids, and asphalt. The petrochemical products that have the greatest impact on our lives are plastics. These materials have an extremely wide range of physical properties and have replaced metals, glass, and wood in many applications.

The impact of fossil fuels as a convenient and transportable source of energy and as a source of plastics and fertilizers has been enormous. In earlier times, almost half the population was involved in agriculture and lived in rural areas. The advent of powerful machines changed the nature of agriculture and allowed a small number of people to take care of various aspects of producing food crops. Industrial farming is almost always done with synthetic fertilizers that are produced from components of petroleum and natural gas. The easy accessibility of an abundant energy source has changed modern lives in many ways. Personal automobiles and mechanized transportation have allowed people to live far from their place of work, thus greatly expanding the area of human habitation. International trade with ships or airplanes allows people living anywhere to obtain products from distant lands.

From the very beginning of industrialization, it was apparent that a price had to be paid for these radical changes. Workers in many professions were displaced; they often had to find work in assembly lines in factories. The new arrivals to cities often had to live in congested, run-down apartments. Workers who extracted fossil fuels from mines and people living in the neighborhood were invariably exposed to chemicals that were dangerous and harmful for health. Accidents in coal mines that kill many miners take place even now. Combustion of fossil fuels, the process by which energy was extracted from them, invariably releases dangerous effluents into the atmosphere.

Coal

Among the fossil fuels, coal has been known to humans for a very long time as a condensed source of energy. Archaeological research suggests that coal was used in funeral pyres about 3,000 years ago. Aristotle mentions coal as the rock that burns. Archaeologists have also found evidence that Romans used coal to heat their homes in the 2nd century AD and Hopi Indians in the U.S. Southwest used coal for cooking, heating, and baking their exquisite pottery. The early machines of the Industrial Revolution, including steam locomotives, steel mills, and textile factories, depended on the use of coal because its energy content was much greater than that of wood.

Coal played a major role in the early stages of industrialization and is still widely used for the generation of electricity. Early steam engines and other machines worked by using energy from coal. It is still a major fuel to produce energy—both in the United States and in almost all other countries.

Coal-fired power plants use an enormous amount of coal to produce electricity. For example, on a hot summer day the giant Gibson station in southwestern Indiana consumes 25 tons of coal per minute to produce 3,000 megawatts (MW) of electric power.[358] A modest power station that produces 500 megawatts of electricity still uses about 4,000 tons of coal per day.

The importance of coal can be judged from the fact that it provides roughly 40 percent of the world's electricity these days. There are approximately 2,400 coal-fired power plants in the world. Their number is fluid because some developing countries, particularly China and India, are building new power plants while developed countries are retiring smaller plants. In 2019, the worldwide consumption of coal was about 9 billion tons. China is the biggest producer and consumer of coal and used almost half the world's production of coal (4.7 billion tons) to generate 3.25 trillion kilowatt-hours (KWH) of electricity, 80 percent of its energy requirements.[359] The United States produced 2 billion KWH of electricity from coal-fired power plants, representing 40 percent of the energy produced in the country. India, the third largest producer and consumer of coal, produced 644 million KWH of electricity, representing 59 percent of the total energy produced in the country.[360] China's construction firms are also building coal-fired power stations in other countries.

The carbon content of coal depends on the age of the rock, i.e. the period for which it has remained buried in the Earth, and it slowly increases with the passage of time. Since the efficiency in producing heat depends on the amount of carbon in it, rocks that have remained buried for longer periods are more efficient and make better fuel sources because they produce more energy per unit weight of the rock. The types of coal in order of increasing carbon content are lignite, sub-bituminous, bituminous, and anthracite. In almost all coal plants, coal is pulverized into fine particles and then burned to produce steam, which is then passed through a turbine to produce electricity. Combustion of carbon in coal or other fossil fuels in power plants produces enormous amount of carbon dioxide—1 kilogram of carbon produces about 3.6 kilograms of gaseous carbon dioxide. Since coal-based power plants combust an enormous amount of coal, the amount of this gas released in the atmosphere is very large. Carbon dioxide is partially soluble in water, hence the gas dissolved in the water of oceans makes it more acidic.

Coal creates serious environmental and human health problems in all stages of production and consumption. There are two common methods of mining coal—strip mining and underground mining. In strip mining, the top of a mountain is removed to expose the coal. This process has serious environmental impacts, including the destruction of forests, landscapes, and wildlife habitats. Rain washes the loosened soil into waterways and its chemical contaminants seep into the groundwater, thereby contaminating it. Strip mining in the Appalachian range has destroyed the pristine forests of West Virginia, Kentucky, Virginia, and Tennessee and converted those regions into barren and polluted lands. In China and the United States, the two countries that burn the greatest amount of coal, millions of hectares of land have been permanently degraded. Reclamation of the land, often promised by the industry and the governments, is very difficult because of the loss of topsoil and the presence of coal dust and other contaminants in the soil.

When coal is buried deep inside the Earth, it is excavated by underground mining. Often the removal of vast amounts of coal leads to subsidence in which the land collapses in the areas above the mines. Underground mining brings large amounts of waste to the surface, which may contain toxic substances. The empty space created by the mining operations changes the flow of groundwater and streams. Coal mining is a dangerous occupation that employs about 120,000 workers in the United States. Physical hazards in mining include injuries from falling rocks, fires, explosions, mobile equipment accidents, entrapment, and electrocution. The number of fatalities due to accidents in coal mines has decreased substantially in the United States due to better regulations but about 1,000 people die in mines around the world each year in underground coal-mining operations.[361] While fatalities are caused by accidents, work in coal mines itself is dangerous. Studies have shown the lives of coal miners and those living around coal mines are shortened due to black lung disease, chronic bronchitis, and emphysema.[362] Prolonged exposure to coal dust often causes pneumoconiosis, or "black lung" and chronic obstructive pulmonary disease. The U.S. National Institute of Occupational Safety and Health reported that close to 9 percent of the miners with twenty-five years of experience tested positive for black lung.[363] Cumulative effects of this disease kills about 1,000 former coal miners in the United States each year.[364] Mining of coal is also harmful for the ecology. During the period between 1985 and 2001, more than 7 percent of the Appalachian Forest was cut down and 1,200 miles of streams were buried or polluted due to underground mining.

Greenhouse Effect: The combustion of all fossil fuels produces carbon dioxide in large amounts. However, the combustion of coal produces more carbon dioxide per unit of energy than petroleum or natural gas because the hydrogen in natural gas and petroleum also produces water on combustion. The emission of carbon dioxide, a greenhouse gas, increases every year. The 2,425 coal-fired power plants in the world produced 2,000 GW of electricity and emitted 15 billion tons of carbon dioxide in 2019. The 359 operating coal plants in the United States emitted 1.7 billion tons of gaseous carbon dioxide into the atmosphere. Coal plants also emit various oxides of nitrogen which, though much smaller in amount, stay in the atmosphere for longer periods. Nitrogen oxides and methane are more potent greenhouse gases than carbon dioxide because they are more efficient in trapping the heat reflected from the Earth and stay in the atmosphere for longer periods.

Air Pollution and Acid Rain: Burning coal is the leading cause of smog, acid rain, and air pollution. Nitrogen oxide gases produced during the combustion of coal generate ground level ozone, which produces smog in the lower atmosphere. Ozone exposure exacerbates asthma and causes respiratory diseases. Combustion of fossil fuels also produces gaseous sulfur dioxide. In the United States, roughly two-thirds of all sulfur dioxide and one-quarter of all nitrogen oxides originate from power plants that produce electricity from coal. Sulfur dioxide makes the lungs of people sting and also damages the exterior of buildings. Acid rain is formed when gases emitted during the combustion of fossil fuels react with oxygen, water vapor, and other chemicals. It creates serious environmental problems because the increased acidity in water adversely affects the survival of some marine species. Acid rain also makes it difficult for plants to absorb minerals that they require for healthy growth, thereby stunting the growth of vegetation.

Coal Ash and Particulate Matter: Coal plants produce large amount of slurry as a residual material known as coal ash, which contains toxic substances such as arsenic, chromium, lead, and mercury. The power plants in the U.S. generate 110 million tons of this waste product, which gets mixed up with water and stored in 'coal ash ponds'. This coal slurry is very toxic and dangerous for humans. Short-term exposure to this toxic mix causes irritation of the nose and throat, dizziness, and vomiting, and long-term exposure causes liver and kidney damage, cardiac arrhythmia, and a variety of cancers. Combustion of coal also produces particulate matter, appearing in the form of haze in the vicinity of coal-fired power plants. PM2.5 particles, less than 2.5 microns in diameter, are of particular concern because they are able to travel deep into the respiratory tract and reach the lungs, causing respiratory and cardiovascular problems. Airborne particulate matter can cause bronchitis and asthma and may result in premature death.

Toxic Substances: Coal often contains arsenic, cadmium, boron, lead, and other toxic metals that have been shown to be injurious to human health. Even though the proportions of these impurities in coal are small, the amount of coal combusted in power plants is so large that they have a substantial cumulative effect on the environment and human health. Arsenic causes nervous system damage, cardiovascular issues, and urinary tract cancers. Lead has been implicated in kidney diseases and cardiovascular problems, and boron adversely affects human vision. The Environmental Protection Agency has found that living next to a coal ash disposal site can increase the risk of cancer and other diseases.

Mercury: Out of all toxic substances, mercury merits special mention because its effects are not confined to the vicinity of power plants. In the United States, it is estimated that 80 tons of mercury is released into the atmosphere in the form of vapors where it combines with methane to form a compound, methylmercury.[365] It is carried to distant lands by wind and comes down to the ground level with rains that eventually carry it to the sea. Algae absorb methylmercury and thus it enters the food chain of the aquatic animals. Since it is not excreted by marine organisms, its concentration keeps increasing as it moves higher up the food chain. After ingestion of contaminated fish by humans, mercury causes nervous system damage with the greatest risk to infants and fetuses.

Radiation: A series of studies have shown that fly ash emitted by coal power plants carries one hundred times more radiation into the surrounding environment than a nuclear power plant. Combustion of coal removes carbon and some other contaminants so that the concentration of radioactive substances such as uranium and thorium increases.[366]

Water Consumption and Pollution: A coal-fired power plant requires much more water than a power plant using other fuels. Coal is usually washed with water to remove sulfur and other impurities so that the emission of ash and sulfur dioxide during the combustion process is reduced. Like all thermoelectric power systems, coal plants require water for the cooling systems that is usually withdrawn from nearby bodies of water, such as lakes, rivers, and seas. The coal ash or sludge produced by scrubbers is stored in open-air pits from which it often leaches into groundwater. According to the EPA, coal-fired power plants are the nation's biggest water polluters because more than 50 percent of all toxic water pollution originates from them.

Prospects: In 2014, the electric power industry unveiled a plan to retire more than 10 percent of the coal-fired power plants in the United States within a decade. However, the 140 power plants that are slated for retirement are mostly small and old plants in the Midwest and South, which will decrease carbon dioxide emissions by only about 4 percent.[367] New emission regulations imposed by the EPA, as well as low natural gas prices, may force the industry to shut down a few other coal-fired power plants. Even so, coal is projected to remain a significant source of electricity in the United States in the coming decades. The International Energy Agency predicts that the global coal demand will keep rising at the rate of about 2 to 3 percent, driven mostly by increasing demand for power in developing countries.

Coal is a popular fuel because it is available in almost all countries in the world. Its large proven reserves of 861 billion tons worldwide will last for 122 years at the current rate of consumption. For comparison, the proven oils and gas reserves are projected to last for forty-six and fifty-four years, respectively. Coal is also relatively easy to obtain from mines, which makes it a cheap fuel. The price of electricity obtained from coal-fired plants is low—but only if one ignores the lasting cost of air and water pollution, the obliteration of landscapes, and the endangerment of human health in various phases of coal's production and combustion. But perhaps the most important adverse effect of the use of coal is the production of copious amounts of carbon dioxide, which has deleterious effect on the global climate.

Crude Oil

The discovery of petroleum added another dimension to the applications of fossil fuels because it could be easily transported and ignited when desired. In addition, industrial chemists quickly discovered that components of crude oil can be used to make a large number of petrochemical products. The variety of these products is very large and they have become so ubiquitous that we do not even realize their relation to crude oil. Items made from petrochemicals include numerous household items such as plastics, soaps, solvents, clothing, drugs, fertilizers, pesticides, synthetic fibers, and rubbers. The components of oil extracted from the ground are separated in a refinery by the process of fractional distillation, which separates them according to their boiling points as shown in Table 1. In general, one barrel (42 gallons) of crude oil produces 20 gallons of gasoline for motors, 10 gallons of diesel, 4 gallons of jet fuel, and 15 gallons of other materials that include naphtha, kerosene, liquefied refinery gas, etc. Plastics are made from some of these distillation products, which

Table 1 Fractional Distillation of Crude Oil.

Temperature	Product	Usage
20°C	Refinery gas	Bottled and used as fuel
30°C	Petrol or gasoline	Fuel used in cars
100°C	Naphtha	Used to make chemicals
170°C	Kerosene	Fuel for jet planes
260°C	Diesel	Fuel for diesel engines
280°C	Lubricating oil	Oil for machines
300°C	Fuel oil	Fuel for ships and factories
340°C	Residue	Used to make asphalt

include liquid petroleum gas (LPG) and natural gas liquids (NGL), but a significant portion of plastics is also made from natural gas.

Gasoline and other derivatives of crude oil have an enormous advantage over coal because they can be carried in containers and ignited to produce power whenever needed. For this reason, gasoline, diesel, and kerosene play an important role in the transportation sector, which moves people and goods by cars, trucks, trains, ships, airplanes, or other vehicles. Cars and light trucks contribute about 62 percent of the total carbon dioxide produced by the transportation sector and nearly one-fifth of the country-wide emissions. The contribution of every gallon of gas to the production of greenhouse gases includes 5 pounds during the extraction, production, and delivery of the fuel, and 19 pounds during the ignition that provides power to the vehicles.[368] The U.S. transportation sector makes a greater contribution to the greenhouse gases than that of most other countries because of a much greater number of vehicles. Although CO_2 is the main component in the emissions from the fuels used in the transportation sector, about 3 percent consists of HFCs (hydrofluorocarbons). The global warming potential of HFCs is much greater than that of carbon dioxide—anywhere from a few hundred to a few thousand times that of CO_2, depending on the exact molecular structure.[369]

Plastics: Plastics, which are manufactured from various distillation products of crude oil, have become ubiquitous in our daily lives. Although the contribution of plastics to the emission of greenhouse gases is small, the fact that they do not easily degrade makes them a major source of pollution, which is beginning to have a major impact on the ecosystem and wildlife. In the United States, plastics are manufactured from petrochemical products, including liquid petroleum gases (LPG) and natural gas liquids (NGL). Plastics are polymers made by chemically combining some simple molecules into two- or three-dimensional structures that are very stable. Industrial chemists have synthesized a very large number of plastics with widely different chemical and physical properties.

In 2010, about 191 million gallons of LPG and NGL and 214 billion cubic feet of natural gas were used to make plastic products in the U.S. Because of the low cost of manufacturing plastic products, they are quickly disposed of and replaced with new products. As a result, 280 million metric tons of plastics are thrown away every year throughout the world. Over 1 trillion plastic bags are used every year, which is over 1 million plastic bags every minute. It is estimated that Americans throw away 300 trillion plastic bottles and use 100 billion single-use plastic bags each year. Due to the stability of their structure, most plastics will not biodegrade for centuries. Plastic bags are the second most common type of ocean refuse, second only to cigarette butts. Every square mile of ocean has about 46,000 pieces of plastic floating in it. According to the U.S. National Oceanic and Atmospheric Association (NOAA), plastics floating in seas and oceans cause the death of 1 million sea birds and 100,000 mammals each year and ocean plastic causes around $13 billion of damage every year.

In addition to creating environmental problems, plastics also have an adverse effect on human health. Bisphenol-A (BPA) is often added to plastics to make them durable. It was originally given to cows and chickens to increase their weight before slaughter and is known to disrupt hormones in the body. BPA can mimic the effect of estrogen in the body, leading to weight gain and hormone imbalance. It has also been linked to many health problems, including breast and prostate cancers and type 2 diabetes. The Center of Disease Control (CDC) estimates that 92 percent people in the country have BPA in their bodies. Recent studies have shown that BPA is transformed in the human body into a compound that may cause obesity.[370] Phthalates are also found in many plastics and are often present in indoor air. Phthalates are linked with immune system impairment, reduced testosterone, and many other problems. The European Union banned their use in 2005.

Natural Gas

Natural gas, like all fossil fuels, is a product of chemical reactions in some organisms. During a period of thousands of years, the intense heat and high pressure caused the life forms to release a gas consisting of hydrocarbons, which was then locked in deep underground rock formations. It is usually found in close proximity to crude oil. Natural gas consists mainly of methane but is often mixed with heavier hydrocarbons, such as ethane, butane, and pentane. Some of these impurities are removed during the purification process before the natural gas is released for use in households. Another source of natural gas that has recently come to prominence is shale, a sedimentary rock formed from the compaction of silt and clay-size mineral particles. Shale rocks have different compositions. Black shale rocks obtain their color from tiny particles of organic matter that were deposited when the shale was formed. Over a period of hundreds of thousands of years, the heat in the earth and compaction of the rocks under high pressure transformed the organic matter into oil and natural gas, which migrated out of the shale and were trapped within the pores of overlying rocks.

A method of extracting fossil fuels from shale rocks has been developed in the last decade or so that combines horizontal drilling with hydraulic fracturing and is known as fracking. A 2-kilometer deep well is dug, and the surrounding shale is then drilled in the horizontal direction for about a kilometer. Millions of gallons of fracking fluid are subsequently injected into fissures in the rocks at a high pressure. The fracking fluid consists of water and various chemicals; the exact composition of the fracking fluid is a secret guarded by the industry. The injection of fracking fluid causes the fossil fuels to flow to the production well.

The fracking industry is rapidly increasing in the U.S. and hundreds of thousands of wells have been dug all across the country. Fracking has been most productive in the Marcellus shale formation which stretches across West Virginia, Pennsylvania, and New York. Companies have dug more than 8,000 wells, and about one hundred more are dug every month.[371] Three other major shell "plays" are Barnett in Texas, the Fayetteville in Arkansas, and the Haynesville, which straddles the Texas-Louisiana border. Advanced fracking methods pollute the water, air, and land, and endanger the health of people living in the neighborhood of these facilities. Oil and gas production has been linked to an increased risk of cancer and birth defects. These operations also have an enormous impact on water in those regions because fracking requires millions of gallons of water. Most of the wastewater returned from these operations is polluted with chemicals and is not suitable for drinking.[372]

For some time, people living in the neighborhood of fracking wells have suspected that their operation increases the frequency of earthquakes. A report released by the U.S. Geological Survey in April 2015 determined that fracking operations have led to dramatic increase in the number of earthquakes, primarily from the injection of the wastewater byproducts into underground wells. This process can activate dormant faults in the underground rocks.[373] Eight states in the Eastern and Central U.S. in which fracking operations were done have had dramatic increases in the number and intensities of earthquakes. However, the powerful oil and natural gas industry has ignored the deleterious effects of their operations on the local environment and health of communities, including contaminated water supplies, dangerous air pollution, the destruction of streams, and the devastation of landscapes.

NOTES

Chapter 1: Climate Change

1 "International Energy Outlook 2017," US Energy Information Administration, September 14, 2017.
2 "Frequently Asked Questions," U.S. Energy Information Administration, April 4, 2018.
3 "World Energy Outlook 2017," International Energy Agency.
4 "Climate Change 2013: The Physical Science Basis," Intergovernmental Panel on Climate Change.
5 "Understanding Global Warming Potentials," Greenhouse Gas Emissions, EPA.
6 "Climate Change 2007: Working Group I: The Physical Science Basis," Intergovernmental Panel on Climate Change.
7 "Overview of Greenhouse Gases," EPA.
8 "Total Greenhouse Gas Emissions (kt of CO_2 Equivalent)," The World Bank.
9 Hollie Riebeek, "Paleoclimatology: The Oxygen Balance," NASA Earth Observatory, May 6, 2005.
10 Jouzel et al., "Orbital and Millennial Climate Variability Over the Past 800,000 Years," *Science*, 317 (5839), 793–797, 2007.
11 "Global Climate Report, April 2016," National Centers for Environmental Information, National Oceanic and Atmospheric Administration.
12 Volker Quaschning, "Specific Carbon Dioxide Emissions of Various Fuels," https://www.volker-quaschning.de/datserv/CO2-spez/index_e.php.
13 http://data.giss.nasa.gov/gistemp/graphs_v3/Fig.A2.gif.
14 D. Lyon, "EPA Draft Says Oil & Gas Methane Emissions Are 27 Percent Higher than Earlier Estimates," Environmental Defense Fund, February 23, 2016.

15 R. Howarth, "Methane Emission and Climate Warming Risk from Hydraulic Fracturing and Shale Gas Development: Implications for Policy," Energy Emission Control Technologies, October 8, 2015.

16 "A Dirty Little Secret: Natural Gas's Reputation as a Cleaner Fuel than Coal and Oil Risks Being Sullied by Methane Emissions," *The Economist*, July 23, 2016.

17 J. Hansen. et al., "Target Atmospheric CO_2: Where Should Humanity Aim?" *Cornell University Library*, October 15, 2008.

18 "How Do We Know That Recent CO_2 Increases Are Due to Human Activity?" *Real Climate*, December 22, 2004.

19 "Scientific Consensus: Earth's Climate Is Warming," NASA, *Global Climate Change*.

20 John Cook et al., "Quantifying the Consensus on Anthropogenic Global Warming in the Scientific Literature," *Environmental Research Letters*, Volume 8, Number 2, May 15, 2013.

21 B. Rasmus et al., "Learning from Mistakes in Climate Research," *Theoretical and Applied Climatology*, Volume 126, Issue 3–4 (2016): 609–703.

22 James Powell, "Climate Scientists Virtually Unanimous," *Bulletin of Science, Technology & Society*, March 28, 2016.

23 David Eckstein, Marie-Lena Hutfilz, and Mark Winger, "Global Climate Risk Index 2019," *German Watch.org*, Briefing Paper.

24 N.S. Diffenbaugh, D.L. Swain et al., "Anthropogenic Warming has Increased Drought Risk in California," *Proceedings of the National Academy of Sciences*, March 31, 2015.

25 Brian Howard, "Worst Drought in 1,000 Years Predicted for American West," *National Geographic*, February 12, 2015.

26 R. Gonzales, "California Appears Headed Back to Drought," *NPR*, February 2, 2018.

27 "Hurricanes and Climate Change," Center for Climate Change and Energy Solutions.

28 J. Bacon, "Hurricane Maria Killed More than 4,600 People—More than 70 Times the Official Toll of 64," *USA Today*, May 29, 2018.

29 B.S. Son Kim et al., "Weakening of the Stratospheric Polar Vortex by Arctic Sea-Ice Loss," *Nature Communications*, 5:4646, September 2, 2014.

30 A. Thompson, "The Climate Context for India's Deadly Heat Wave," *Scientific American*, June 4, 2015.

31 J. Worland, "How Climate Change Is Making Wildfires Worse," *TIME*, July 15, 2015.

32 Matt Vilano, "Northern California's Wine Country Needs You to Visit," *CNN*, November 9, 2017.

33 Meg Wagner, Brian Ries, and Veronica Rocha, "Deadly Wildfires Burn in California," *CNN*, July 30, 2018.

34 Edward Helmore, "Unprecedented: More than 100 Arctic Wildfires Burn in Worst Ever Season," *Guardian*, July 19, 2019.

35 Dan Mitchell et al., "The Day the 2003 European Heatwave Record Was Broken," *The Lancet*, July 8, 2019.

36 Stephen Leahy, "Hidden Costs of Climate Change Running Hundreds of Billions a Year," *National Geographic*, September 27, 2017.

37 Alex Morales, "Typhoon Fuels Calls for Global Warming Compensation Funds," *Bloomberg*, November 18, 2013.

38 U.S. Energy Information Administration.

39 Evan Halper, "Fact Check: Trump Makes an Odd Remark About 'Beautiful, Clean Coal,'" *LA Times,* January 30, 2018.

40 "Coal's Role in Electricity Generation Worldwide," World Coal Association.

41 "Highest CO_2 Emitting Power Plants in the World," Carbon Monitoring for Action.

42 Julia Pyper, "Cars Will Cook the Planet Absent Shift to Public Transportation," *Scientific American*, September 14, 2014.

43 Anand Saxena, *The Vegetarian Imperative*, (Baltimore: Johns Hopkins University Press, 2011), 3.

44 Beth Hoffman, "How Increased Meat Consumption in China Changes Landscapes Across the Globe," *Forbes*, March 26, 2014.

45 Saxena, *The Vegetarian Imperative*, 9.

46 C. Foster et al., "Environmental Impacts of Food Production and Consumption: A Report to the Department of Environment, Food, and Rural Affairs," Manchester Business School, Defra, London, 2006.

47 Akifumi Ogina et al., "Evaluating Environmental Impacts of the Beef Cow Calf System by the Lifecycle Assessment Method," *Animal Science Journal*, Volume 78, No. 4 (August 2007): 424–432.

48 Bob Bailey, Antony Froggatt, and Laura Wellesley, "Livestock—Climate Change's Forgotten Sector" (Chatham House, London: The Royal Institute of International Affairs, December 3, 2014).

49 "Rearing Cattle Produces More Greenhouse Gases Than Driving Cars, UN Report Warns," *UN News*, November 29, 2006.

50 *Ibid*.

51 Gidon Eschel et al., "Land, Irrigation Water, Greenhouse Gas, and Reactive Nitrogen Burdens on Meat, Eggs, and Dairy Production in the United States," *Proceedings of the National Academy of Sciences*, 111(33) (2014): 11,996–12,001.

52 T.R. Karl, J.M. Melillo, and T.C. Peterson (eds.), *United States Global Change Research Program*. (New York: Cambridge University Press, 2009).

53 Suzanne Goldenberg, "Climate change 'Already Affecting Food Supply'—UN," *The Guardian*, March 30, 2014.

54 "The Human Cost of Weather Related Events Disasters, 1995–2015," United Nations Office of Disaster Reduction, UNISDR, November 23, 2015.

55 "Climate Change and Health," World Health Organization, February 1, 2018.

56 Erin Thead, "Sea Level Rise: Risk and Resilience in Coastal Cities," Climate Institute, October 2016.

57 Chris Mooney, "Why U.S. East Coast Could Be a Major 'Hotspot' for Rising Seas," *The Washington Post*, February 1, 2016.

58 Helen Briggs, "U.S. Sea Level North of New York City 'Jumped by 128 mm'," *BBC News*, February 24, 2015.

59 "Stern Review on the Economics of Climate Change," http://webarchive.nationalarchives.gov.uk/+/http:/www.hm-treasury.gov.uk/sternreview_index.htm.

60 Thomas Wagner, "Cities at Risk of Rising Sea Levels," *The Associated Press*, March 27, 2007.

61 Stephane Hellegate et al., "Future Flood Losses in Major Coastal Cities," *Nature Climate Change* 3 (2013): 802–806.

62 John Vidal, "Most Glaciers in Mount Everest Region Will Disappear with Climate Change—Study," *The Guardian*, May 22, 2015.

63 Anup Shah, "Climate Change Affects Biodiversity," *Global Issues*, January 19, 2014.

64 Patrick Lynch, "Secrets from the Past Point to Rapid Climate Change in the Future," NASA Global Climate Change, December 14, 2011.

65 Arthur Neslen, "Carbon Dioxide Levels in Atmosphere Forecast to Shatter Milestone," *The Guardian*, March 13, 2016.

66 Andrea Thompson, "Climate Scientists: 2 Degrees of Warming Too Much," *Live Science*, December 4, 2013.

67 "The Low Carbon Economy Index 2017," *PWC*, U.K.

68 Somini Sengupta and Lisa Friedman, "At U.N. Climate Summit, a Call for Action Yields Few Commitments" *New York Times*, September 24, 2019.

69 Julia Pyper, "Cars Will Cook the Planet Absent Shift to Public Transportation," *Scientific American*, September 17, 2014.

70 *Ibid.*

71 *Ibid.*

72 Andy Reisinger and Harry Clark, "How Much Do Direct Livestock Emissions Actually Contribute to Global Warming?" *Global Change Biology*, Volume 24, Issue 4 (April 2018): 1,749–1,761.

73 Saxena, *The Vegetarian Imperative*, 62.

74 Marcin Baranski et al., "Higher Antioxidant and Lower Cadmium Concentrations and Lower Incidence of Pesticide Residues in Organically Grown Crops: A Systematic Literature Review and Meta-Analyses," *British Journal of Nutrition*, Volume 112 (September 2014): 794–811.

75 United Nations Conference on Trade and Development (UNCTAD) and the World Trade Organization (WTO), Technical Paper, "Organic Farming and Climate Change," 2007.

76 Damian Carrington, "IPCC Climate Change Report: Averting Catastrophe Is Eminently Affordable," *The Guardian*, April 13, 2014.

77 Intergovernmental Panel on Climate Change, "Climate Change 2013: The Physical Science Basis."

78 Stern Review, "The Economics of Climate Change," http://mudancasclimaticas. cptec.inpe.br/~rmclima/pdfs/destaques/sternreview_report_complete.pdf.

79 Ian Perry, "Carbon Pricing: Good for You, Good for the Planet," *IMF Blog*, September 17, 2014.

80 Jacob Poushter and Dorothy Manevich, "Globally, People Point to ISIS and Climate Change as Leading Security Threats," Pew Research Center, August 1, 2017.

81 Richard Wike, "What the World Thinks About Climate Change in 7 Charts," Pew Research Center, April 18, 2010.

82 Ronald Brownstein, "Why Republicans Are Frozen in Climate Change," *CNN*, September 12, 2017.

83 Alex McKechnie, "Not Just Koch Brothers: New Drexel Study Reveals Funders Behind the Climate Change Denial Effort," Drexel University, December 20, 2013.

84 Douglas Fischer, "'Dark Money' Funds Climate Change Denial Effort," *Scientific American*, December 23, 2013.

85 Suzanne Goldenberg, "Work of Prominent Climate Change Denier Was Funded by Energy Industry," *The Guardian*, February 11, 2015.

86 Justin Gillis and John Schwartz, "Deeper Ties to Corporate Cash and Doubtful Climate Researcher," *New York Times*, February 22, 2015.

87 Suzanne Goldenberg and Helena Bengtsson, "America's Biggest US Coal Company Backed Dozens of Climate Change Deniers," *Newsweek*, June 15, 2016.

88 Dana Nuccitelli, "Conservative Media Outlets Found Guilty of Biased Global Warming Coverage," *The Guardian*, October 11, 2013.

89 Jeffrey Kluger, "The Climate Denier's Newest Argument," *TIME*, September 29, 2014.

90 Paul Voosen, "NASA Cancels Carbon Monitoring Research Program," *Science*, Volume 360, Issue 6389 (2018): 586–588.

91 Sarah Kaplan and Emily Guskin, "Climate Change Scares and Angers Most U.S. Teens," *Washington Post*, September 16, 2019.

92 Bill McKibben, "The Movement to Divest from Fossil Fuels Gains Momentum," *The New Yorker*, December 21, 2017.

93 "Organic Farming Profits," *Wealth Daily*, August 15, 2013.

94 "Today in Energy: Consumption/Demand," U.S. Energy Information Administration.

95 Naomi Klein, *This Changes Everything: Capitalism vs. the Climate* (New York: Simon and Schuster Paperback, August 2015), 91.

Chapter 2: Water: The Most Precious Resource

96 Ellie Kincaid, "California Isn't the Only State with Water Problems," *Business Insider*, April 21, 2015.

97 "The Quality of Our Nation's Waters," EPA.

98 "Water for a Sustainable World," UN World Water Development Report.

99 Robin McKie, "Why Freshwater Shortages Will Cause the Next Great Global Crisis," *The Guardian*, March 7, 2015.

100 D. Seckler, R. Barker, and U. Amarsinghe, "Water Scarcity in the Twenty-First Century," *International Journal of Water Resource Development* 15 (1999): 29–42.

101 Suzanne Goldenberg, "Global Water Shortages to Deliver Severe Hits to Economies, World Bank Warns," *The Guardian*, May 3, 2016.

102 Mekonnen and Hoekstra, "Four Billion People Facing Severe Water Scarcity," *Science Advances*, February 12, 2016.

103 Elaine Povich, "40 States Expected to See Water Shortages in the Next Decade," *Stateline*, April 16, 2015.

104 *Ibid*.

105 Sarah Zielinski, "The Colorado River Runs Dry," *Smithsonian Magazine*, October 2010.

106 Sydney Weil, "How Does Water Use in the United States Compare to That in Africa," *American Wildlife Foundation*, August 3, 2013.

107 "Crop Water Needs," UN-FAO.

108 "Thirsty Crops: Our Food and Clothes: Eating Up Nature and Wearing Out the Environment?" WWF, January 15, 2013.

109 Silva de Miranda et al., "Environmental Impacts of Rice Cultivation," *American J. of Plant Sciences*, 6 (2015): 2,009–2,018.

110 K. Averyt et al., "Freshwater Use by U.S. Power Plants: Electricity's Thirst for Precious Resource," *A Report of the Energy and Water in a Warming World Initiative*, Union of Concerned Scientists, 2011.

111 "Study: Third of Big Groundwater Basins in Distress," NASA, June 16, 2015.

112 "GroundwaterDepletion,"USGS,https://pubs.er.usgs.gov/publication/70140758/.

113 David Steward et al., "Tapping Unsustainable Groundwater Stores for Agricultural Production in the High Plains Aquifer of Kansas, Projections to 2110" *Proceedings of the National Academy of Sciences of the United States*, Volume 110, No. 37 (2014): E3477–3487.

114 "Saudi Arabia's Great Thirst," *National Geographic*, https://nationalgeographic.com/environment/freshwater/saudi-arabia-water-use/.

115 Aria Bendix, "7 Places That are Sinking Faster Than Anywhere Else in the U.S.," *Business Insider*, January 8, 2019.

116 "Chemicals Evaluated for Carcinogenic Potential: Annual Cancer Report, 2017," US-EPA.

117 "Atrazine, Fertilizer Targeted by President's Cancer Panel," *No-Till Farmer*, July 1, 2010.

118 Saxena, *The Vegetarian Imperative*, 35.

119 Danielle Sedbrook, "2-4D: The Most Dangerous Pesticide You Have Never Heard Of," National Resources Defense Council, March 15, 2016.

120 James Conca, "It's Final—Corn ethanol Is of No Use," *Forbes*, April 20, 2014.

121 De Fraiture et al., "Biofuels and Implications for Agricultural Water Use," *Water Policy* 10 Supplement 1 (2008): 67–81.

122 Jonathan Foley, "It's Time to Rethink America's Corn System," *Scientific American*, March 2013.

123 Andy Koenig, "Stop Fueling the Corn Lobby's Dirty Ethanol Mandate," *Forbes*, January 25, 2016.

124 A.Y. Hoekstra and A.K. Chapagain, "Water Footprint of Nations: Water Use by People as a Function of Their Consumption Pattern," *Water Resource Management* 21 (2007): 35–48.

125 A.Y. Hoekstra (ed), "Virtual Water Trade," Proceedings of the International Expert Meeting on Virtual Water Trade, IHE Delft, The Netherlands.

126 M. Wackernagel and W. Rees, *Our Ecological Footprint: Reducing Human Impact on the Earth*, (Gabriola Island, BC: New Society Publishers, 1998).

127 A.Y. Hoekstra and P.Q. Hung, "Virtual Water Trade: A Quantification of Virtual Water Flows Between Nations In Relation To International Crop Trade," *Value of Water Research Report Series No. 11*, IHE Delft, The Netherlands, September 2002.

128 A.Y. Hoekstra and A.K. Chapagain, "Water Footprint of Nations," *Water Resource Management*, 2006.

129 Janet Larsen, "Meat Consumption in China Now Double That in the United States," Earth Policy Institute, April 24, 2012.

130 M.M. Mekonnen and A.Y. Hoekstra, "A Global Assessment of Water Footprint of Farm Animal Products," *Ecosystems* (2012): 401–415.

131 A.Y. Hoekstra and P.Q. Hung, "A Quantification of Virtual Water Flows Between Nations in Relation to International Crop Trade," *Value of Water Research Report Series No. 11*, UNESCO-IHE Institute for Water Education, Delft, The Netherlands, 2002.

132 "Growth in U.S. Agricultural Exports to China," USDA Press Release No. 0325-15.

133 U.S. Department of Commerce.

134 Karen Coates, "The Global Land Grab: Rich Countries Are Buying Up Poor Countries," *Slate*, April 25, 2014.

135 John Vida, "Fears for the World's Poor Countries As the Rich Grab Land to Grow Food," *The Guardian*, July 3, 2009.

136 Peter Brabeck Letmathe, "Another Inconvenient Truth," *New York Times*, October 5, 2008; and Fast Check, "Did the Chief Executive Officer of Nestle Say Water Is Not a Human Right?" *Snopes*.

137 Peter Gleick, "The Human Right to Water," *Pacific Institute*, May 2007.

138 Regan Morris, "Nestle: Bottling Water in Drought-Hit California," *BBC News*, May 3, 2016.

Chapter 3: Environmental Degradation

139 "Ecological Footprint," Global Footprint Network.

140 "Earth Overshoot Day," http://www.overshootday.org.

141 *Ibid.*

142 Overshootday.org/newsroom/country-overshoot-days/.

143 Philip Landrgan et al., "The Lancet Commission on Pollution and Health," *The Lancet*, October 19, 2017.

144 "Cars, Trucks, Buses and Air Pollution," Union of Concerned Scientists.

145 "National Air Quality: Status and Trends of Key Air Pollutants," EPA, July 31, 2018.

146 "State of the Air: 2014," American Lung Association.

147 "7 Million Premature Deaths Linked to Air Pollution," World Health Organization, March 26, 2014.

148 Christina Procopiou, "Air Pollution Claims 5.5 Million Lives a Year, Making It the Fourth Leading Cause of Death Worldwide," *Newsweek*, February 12, 2016.

149 "Most China Cities Fail to Meet Air Quality Standards," *BBC News*, February 3, 2015.

150 Eleanor Albert and Beina Xu, "China's Environmental Crisis," *Council on Foreign Relations*, January 18, 2016.

151 Mike Ives, "Pollution May Dim Thinking Skills, Study in China Suggests," *New York Times,* August 29, 2018.

152 "India's Polluted Air Claimed 1.24 Million Lives in 2017: Study," *Reuters,* December 6, 2018.

153 Umair Irfan, "Why Is India's Air Pollution so Horrendous," *VOX,* June 9, 2018. "India's Polluted Air Claimed 1.24 Million Lives in 2017: Study," *Lancet Planetary Health* (2018).

154 "India Cities Dominate World Air Pollution List," *BBC News,* May 2, 2018,

155 "The Cost of Air Pollution," OECD, May 21, 2014.

156 "State of the Air, 2017," American Lung Association.

157 S. Ghude et al., "Reduction in India's Crop Yield Due to Ozone," *Geophysical Research Letters,* August 14, 2014.

158 Jennifer Burney and V. Ramanathan, "Recent Climate and Air Pollution Impacts on Indian Agriculture," *Proceedings of the National Academy of Sciences*, Volume 111, No. 46 (2014): 16,319–16,324.

159 *United Nations Yearbook 2014.*

160 "Health and Environmental Effects of Particulate Matter (PM)," EPA, June 20, 2018.

161 H. Regan, "21 of the World's 30 Cities with the Worst Air Pollution Are in India," *CNN,* February 26, 2020.

162 Peng Yin et al., "China and Its Provinces 1990–2017: An Analysis for the Global Burden of Disease," *The Lancet Planetary Health*, August 17, 2020.

163 Samantha Jakuboski, "The Impact of Asian Air Pollution on the World's Weather," *Nature,* July 1, 2014.

164 Saxena, *The Vegetarian Imperative*, 35–36.

165 Jasmine Brown and Karl Schneider, "Industrial Waste Pollutes America's Drinking Water," The Center for Public Integrity, August 17, 2017.

166 Debbie Elliott, "5 Years After BP Oil Spill, Effects Linger and Recovery Is Slow," *National Public Radio*, April 20, 2015.

167 Pasquale Borrelli et al., "An assessment of the Global Impact of 21st Century Land Use Change on Soil Erosion," *Nature Communications*, December 8, 2017.

168 "Land Cover and Change Detection," Satellite Imaging Corporation.

169 Saxena, *The Vegetarian Imperative*, 62.

170 Chris Arsenault, "Only 60 Years of Farming Left if Soil Degradation Continues," *Scientific American*, December 5, 2014.

171 "World Losing 2,000 Hectares of Farm Soil Daily to Salt-Induced Degradation," United Nations University, October 28, 2014.

172 Dominique Patton, "More than 40 Percent of China's Arable Land Degraded," *Reuters*, November 4, 2014.

173 "World Losing 2,000 Hectares of Farm Soil Daily to Salt-Induced Degradation," *United Nations University*, October 28, 2014.

174 "Special Report on Climate Change, Desertification, Land Degradation, Sustainable Land Management, Food Security, and Greenhouse Gas Fluxes in Terrestrial Ecosystems," IPCC, February 2017.

175 "Desertification in the EU," European Court of Auditors, June 2018.

176 Joe McCarthy, "Land Degradation Costs the World $10.6 Trillion Each Year," *Global Citizens*, September 22, 2018.

177 "Forest Degradation Found to Be Underestimated," Woods Hole Research Center, December 18, 2015.

178 Alina Bradford, "Deforestation: Facts, Causes & Effects," *Live Science*, April 3, 2018.

179 Peter Scholtus, "The US Consumes 1,500 Plastic Water Bottles Every Second, a Fact By Watershed," 2009.

180 "Plastic Facts and Statistics," Container Recycling Institute.

181 Laura Parker, "A Whopping 91% of Plastic Isn't Recycled," *National Geographic*, July 19, 2017.

182 D.W. Laist, "Impact of Marine Debris: Entanglement of Marine Life in Marine Debris Including a Comprehensive List of Species with Entanglement and Ingestion Records," in J.M. Coe and D.B. Rogers (eds.), *Marine Debris: Sources, Impacts and Solutions* (New York: Springer-Verlag, 1997), 99–139.

183 R. Geyer, J. Jambeck and K. Law, "Production, Use, and Fate of all Plastics Ever Made," *Science Advances*, July 19, 2017.

184 R.U. Halden, "Plastics and Human Health," *Ann. Rev. Public Health* (April 30, 2010), 179–194.

185 "Global Plastic Production Rises, Recycling Lags," WorldWatch Institute, August 5, 2018.

186 David Cutler and Francesca Dominici, "A Breath of Bad Air: Cost of Trump Environmental Agenda May Lead to 80,000 Extra Deaths Per Decade," *Journal of the American Medical Association*, 119(22) (2018): 2,261–62.

187 Justin Worland, "Air Pollution Is Still Killing People in the United States," *TIME*, June 28, 2017.

Chapter 4: Depletion of Planetary Resources

188 "Threats: Deforestation and Forest Degradation," worldwildlife.org.

189 "Global Forest Resources Assessment 2010," UN-FAO.

190 "Brazil Deforested 10,000 Square Km of Amazon Rainforest in 2019, Up 34% on Year," *Reuters Environment*, June 30, 2020.

191 Gaia Vince, "How the World's Oceans Could Be Running Out of Fish," *BBC,* September 21, 2012.

192 D. Pauly et al., "Toward Sustainability in World Fisheries," *Nature,* 418 (2002): 689–695.

193 D. Pauly et al., "Fishing Down Marine Food Webs," *Science,* 279 (1998): 860–863.

194 Costello et al., "Status and Solutions for the World's Unassessed Fisheries," *Science,* 338 (6106) (October 2012): 517–520.

195 Charles Ebinger, "6 Years from the BP Deepwater Horizon: What We've Learned, and What We Shouldn't Misunderstand," *Brookings,* April 20, 2016.

196 "Dead Zones," Virginia Institute of Marine Sciences.

197 "Hypoxia," National Ocean Service.

198 Breitburg et al., "Declining Oxygen in Global Ocean and Coastal Waters," *Science* (January 5, 2018).

199 "The State of the World Fisheries and Aquaculture," UN-FAO, 2016.

200 Pallav Ghosh, "Omega-3 Oils in Farmed Salmon 'Halve in Five Years'," *BBC News,* October 6, 2016.

201 R. Hites et al., "Global Assessment of Organic Contaminants in Farmed Salmon," *Science* 303 (2004): 226–229.

202 Gaia Vince, "How the World's Oceans Could Be Running Out of Fish," *BBC News,* September 21, 2012.

203 R. Dirzo et al, "Defaunation in the Anthropocene," *Science,* July 25, 2014.

204 Julie Shaw, "Why Is Biodiversity Important?" *Conservation International,* November 15, 2018.

205 Roger Cohn, "Putting a Price Tag on the Real Value of Nature," Yale Environment 360, January 5, 2012.

206 Intergovernmental Science-Policy Platform on Biodiversity and Ecosystem Services, May 6, 2019.

207 https://www.ipbes.net.

208 World Wildlife Fund, "Living Planet Report 2018."

209 George Manboit, "Neonicotinoids Are the New DDT Killing the Natural World," *The Guardian,* August 5, 2013.

210 Stephen Leahy, "One Million Species at Risk of Extinction, UN Report Warns," *National Geographic,* May 6, 2019.

211 World Trade Organization, "World Trade Statistical Review: 2016," https://www.wto.org/english/res_e/statis_e/wts2016_e/wts2016_e.pdf.

212 Branco Milanovic, "The Real Winners and Losers of Globalization," https://www.theglobalist.com/the-real-winners-and-losers-of-globalization.

213 Rupert Neate, "Ten Billionaires Reap $400 Bn Boost to Wealth During the Pandemic," *The Guardian,* Dec 19, 2020.

214 Kimberly Amadeo, "U.S. Trade Deficit with China and Why It's So High," *The Balance*, June 14, 2018.

215 Will Kimball and Robert E. Scott, "China Trade, Outsourcing and Jobs," Economic Policy Institute, December 11, 2014.

216 John McQuaid, "The Secrets Behind Your Flowers," *Smithsonian Magazine*, February 2011.

217 Lawrence Mishel and Alyssa Davis, "CEO Pay Has Grown 90 Times Faster Than Typical Worker's Pay Since 1978," Economic Policy Institute, Economic Snapshot, July 1, 2015.

218 Mike Collins, "The Pros and Cons of Globalization," *Forbes*, May 6, 2015.

219 Jeffrey Frankel, "Do Globalization and World Trade Fuel Inequality?" *The Guardian*, January 2, 2018.

Chapter 5: Overconsumption

220 Gus Lubin, "There Is a Staggering Conspiracy Behind the Rise of Consumer Culture," http://www.businessinsider.com/birth-of-consumer-culture-2013-2.

221 Herbert Hoover Quotes, http://www.presidential-power.org/quotes-by-presidents/herbert-hoover-quotes.htm.

222 Alejandro Estrada et al., "Impending Extinction Crisis of the World's Primates: Why Primates Matter," *Science Advances,* June 18, 2017.

223 G. Ceballos et al., "Biological Annihilation Via the Ongoing Sixth Mass Extinction Signaled By Vertebrate Population Losses and Declines," *Proc. National Academy of Sciences*, July 25, 2017.

224 Bureau of Economic Analysis, U.S. Department of Commerce, "Personal Income and Outlays," www.bea.gov.

225 "A Look at the Shocking Student Loan Debt: Statistics for 2018," *Student Loan Hero*, updated May 1, 2018.

226 Alexandra Sifferlin "Here Is How Happy Americans are Right Now," *TIME*, July 26, 2017.

227 Thomas DeLeire and Ariel Kahlil, "Does Consumption Buy Happiness? Evidence from the United States," *Int. Rev. Econ.*, 57 (2010): 163–176.

228 "Media Advertising Spending in the United States," The Statistics Portal.

229 "Total US Ad Spending to See Largest Increase Since 2004," eMarketer, July 2, 2014.

230 "The Impact of Food advertising on Childhood Obesity," *American Psychological Association*.

231 Y.C. Wang et al., "Health and Economic Burden of Projected Obesity Trend in the USA and the UK," *The Lancet*, Volume 378 (August 27, 2011): 804–814.

232 Erin El Issa, "2017 American Household Credit Card Debt Study."

233 *Ibid*.

234 Celia Cole, "Overconsumption Is Costing Us the Earth and Human Happiness," *The Guardian*, June 21, 2010.

235 A. Pandey, "US Fossil Fuel Subsidies Increase 'Dramatically' Despite Climate Change Pledge," *International Business Times*, November 12, 2015.

236 "U.S. Apparel Market—Statistics and Facts," Statista.

237 Imran Amed et al., "The State of Fashion 2018," McKinsey & Co.

238 Alden Wicker, "Fast Fashion Is Creating an Environmental Crisis," *Newsweek*, September 1, 2016.

239 Elizabeth L. Cline, "Overdressed: The Shockingly High Cost of Cheap Fashion," *Portfolio*, June 14, 2012.

240 Alden Wicker, "Fast fashion Is Creating an Environmental Crisis," *Newsweek*, September 1, 2016.

241 "Child Labour in the Fashion Supply Chain," UNICEF.

242 "Social and Economic Injustice," WorldCentric.

243 Felicity Lawrence, "How Peru's Wells Are Being Sucked Dry by British Love for Asparagus," *The Guardian*, September 14, 2010.

244 David Martin, "5 Toxics That Are Everywhere: Protect Yourself," *CNN*, May 31, 2010.

245 Edward Wong, "Air Pollution Linked to 1.2 Million Premature Deaths in China," *New York Times*, April 1, 2013.

246 Dave Tilford of Sierra Club, quoted in *Scientific American*, September 14, 2012.

247 "Child Labour," International Labour Organization.

248 Anup Shah, "Corporations and Worker's Rights," Global Issues, May 28, 2006.

249 Mary MacVean "For Many People, Gathering Possessions is Just the Stuff of Life," *LA Times,* March 21, 2014.

250 Martha de Lacey, "Over 600 Dresses, 400 Pairs of Shoes and 1,116 Tops! No, Not Suri Cruise We Mean YOU: Women Will Own 558 Pairs of Trousers and 372 Cardigans Over a Lifetime," *Daily Mail (UK)*, January 10, 2013.

251 Mark J. Perry, "New US homes Today Are 1,000 Square Feet Larger Than in 1973 And Living Space Per Person Has Nearly Doubled," *AE Ideas*, June 5, 2016.

252 Tom Vanderbilt, "Self-Storage Nation," *Slate*, July 18, 2005.

253 "Municipal Solid Waste," EPA.

254 Nadia Kounang, "Dangerous Products Hiding in Everyday Products," *CNN*, July 1, 2016.

255 Amy Westervelt, "Not So Pretty: Women Apply an Average 168 Chemicals Every Day," *The Guardia*n, April 30, 2015.

256 Brian Palmer, "Go West, Garbage Can," *Slate*, February 15, 2011.

257 "America's Food Waste Problem," EPA, April 22, 2016.

258 Elizabeth Royte, "One-Third of Food is Lost or Wasted: What Can be Done?" *National Geographic*, October 2014.

259 "Save Food: Global Initiative on Food Loss and Waste Reduction," UN-FAO.

260 Roni Neff, Marie Spiker, and Patricia Truant, "Wasted Food: U.S. Consumers' Reported Awareness, Attitudes, and Behavior," *PLOS*, June 10, 2015.

261 Roff Smith, "How Reducing Food Waste Could Ease Climate Change," *National Geographic*, January 22, 2015.

262 *Ibid*.

263 Ivan Watson, "China: The Electronic Wastebasket of the World," *CNN*, May 30, 2013.

264 "'E-Waste Pollution' Threat to Human Health," *Institute of Physics*, May 31, 2011,

265 Saxena, *The Vegetarian Imperative*, 31–50.

266 William Rees and Mathis Wackernagel, *Our Ecological Footprint: Reducing Human Impact on Earth* (Gabriola Island, BC: New Society Publishers, 1996).

267 Global Footprint Network, "Ecological Footprint."

268 footprintnetwork.org

269 Rebecca Leber, "This Chart Will Tell You Just How Quickly Your Country is Depleting the Earth," *The New Republic*, September 30, 2014.

270 "Ecological Wealth of Nations," Footprintnetwork.

271 "Sustainable Development Goals: 17 Goals to Transform Our World," United Nations.

Chapter 6: Animal-Based Foods

272 ourworldindata.org/meat-production.

273 Alex Thornton, "This Is How Many Animals We Eat Each Year," *World Economic Forum*, February 8, 2019.

274 Hannah Ritchie, "Meat and Dairy Production," ourworldindata.org/meat-production, November 2019.

275 Alon Sharpton et al., "The Opportunity Cost of Animal-Based Diets Exceeds All Other Food Losses," Proceedings National Academy Sciences (April 10, 2018): 3804–3809.

276 USDA Natural Resources Conservation.

277 David Merill and Lauren Leatherby, "Here's How America Uses Its Land," *Bloomberg*, 2018.

278 Giorgia Guglielmi, "Are Antibiotics Turning Livestock Into Superbug Factories?" *Science*, September 28, 2018.

279 David Wallinga, "Better Burgers: Why It's High Time the U.S. Beef Industry Kicked Its Antibiotics Habit," *NRDC*, June 25, 2020.

280 Ashoka Mukpo, "World Bank's IFC Pumped $1.8 Billion Into Factory Farming Operations Since 2010," *Mongabay Series: Global Commodities*, July 7, 2020.

281 M.M. Mekonnen and A.Y. Hoekstra, "A Global Assessment of the Water Footprint of Farm Animal Products," *Ecosystems*, 15(3) (2012): 401–405.

282 Mekonnen and Hoekstra, "A Global Assessment of the Water Footprint of Farm Animal Products," 401–415.

283 The Guardian Datablog.

284 Food Empowerment Project, Pollution (Water, Air, Chemicals), EPA.

285 "Summary Report on Antimicrobials Sold or Distributed for Use in Food-Producing Animals," FDA.

286 C. Lee Ventola, "The Antibiotic Resistance Crisis," https://bnbi.nlm.nih.gov/pmc/articles/PMC4378521.

287 O'Neil, Jim, "Antimicrobial Resistance: Tackling a Crisis for the Heath and Wealth of Nations," *The Review of Antimicrobial Resistance*, December 2014.

288 Gerber et al. "Tackling Climate Change Through Livestock: A Global Assessment of Emissions and Mitigation Opportunities," FAO, Rome, 2013.

289 Lorraine Chow, "WWF: 60% of Global Biodiversity Loss Due to Land Cleared for Meat-Based Diets," *EcoWatch,* October 05, 2017.

290 Evelyn Battaglia et al. "Health Risks Associated with Meat Consumption: A Review of Epidemiological Studies," *Nutrition Research* 85(1–2), (2015): 70–78.

291 "Risk in Red Meat?" *NIH Research Matters*, March 26, 2012.

292 Roxanne Khamise, "Red Meat Linked to Breast Cancer Risk," *New Scientist*, November 13, 2006.

293 Joan Sabate et al., "Unscrambling the Relation of Egg and Meat Consumption with Type 2 Diabetes Risk," *Am. J. Clin. Nutr.*, Volume 108 (October 2018): 1,121–1,128.

294 John Cumbers, "Preventing Another Pandemic Might Be as Simple as Trying Alternative Meat," *Forbes*, May 9, 2020.

295 Sophie Kevany, "Millions of Farm Animals Culled as US Food Supply Chain Chokes Us," *The Guardian*, April 29, 2020.

296 Ambika Satija and Frank Hu, "Plant-Based Diets and Cardiovascular Health," *Trends in Cardiovascular Medicine*, October 28(7) (2018): 437–441.

297 Joan Sabate, "The Contribution of Vegetarian Diets to Health and Disease: A Paradigm Shift," *American Journal of Clinical Nutrition*, Volume 78, No. 3 (2003): 502S–507S.

298 E.C. Hemler and F.B. Hu, "Plant-Based Diets for Cardiovascular Disease Prevention: All Plant Foods Are Not Equal," *Current Atherosclerosis Rep,* March 20, 2019.

299 Foodprint.org/reports/thefoodprintofeggs/

300 Quanhe Yang et al, "Added Sugar Intake and Cardiovascular Mortality Among US Adults," *JAMA Internal Medicine* 174(4) (2014): 516–524.

301 Boris Worm, et al., "Impacts of Biodiversity Loss in Ocean Ecosystem Services," *Science*, 314 (5800), (2006): 786–790.

Chapter 7: Wealth Concentration

302 DEA: Distributional Financial Accounts, Board of Governors of the Federal Reserve System.

303 Chuck Collins and Josh Hoxie, "New Report Exposes Extreme Concentration of Wealth," Institute of Policy Studies, October 2018.

304 Edward Wolff, "Household Wealth Trends in the United States, 1962 to 2016," National Bureau of Economic Research, Paper No. 24085, November 2017.

305 Zukiewicz-Solczak et al., "Obesity and Poverty Paradox in Developed Countries," *Ann Agric Environ Med* 21(3), (2014): 590–594.

306 James A. Levine, "Poverty and Obesity in the U.S.," *Diabetes* (November 2011): 2,667–2,668.

307 "Top Charts of 2020," Economic Policy Institute, December 28, 2020.

308 David Leonhardt, "The American Dream, Quantified at Last," *New York Times*, December 8, 2016.

309 Raj Chetty et al., "The Fading American Dream: Trends in Absolute Income Mobility Since 1940," *Science*, Volume 356(6336), (April 28, 2017): 398–408.

310 *Ibid.*

311 Yu Xie and Yongai Jin, "Household Wealth in China," *Chin. Sociological. Review*, 47(3), (2015): 203–229.

312 Heather Long, "U.S. Has Lost 5 Million Manufacturing Jobs Since 2000," *CNN*, March 29, 2016.

313 John Komlos, "Why Trade Deals Hurt Americans," *PBS*, June 25, 2015.

314 *Ibid.*

315 Josh Bivens and Lawrence Mishel, "Understanding the Historic Divergence Between Productivity and a Typical Worker's Pay," *Economic Policy Institute*, September 2, 2015.

316 *Ibid.*

317 Elise Gould and Will Kimball, "Right-to-Work States Still Have Lower Wages," *Economic Policy Institute*, April 22, 2015.

318 Mike Collins, "The Decline of Union Is a Middle-Class Problem," *Forbes*, March 19, 2015.

319 Drew Desilver, "10 Facts About American Workers," *Pew Research Center*, September 1, 2016.

320 Florence Jaumotte and Caroline Osorio Buitron, "Power from the People," *Finance and Development* 52, International Monetary Fund, Volume 1, March 2015.

321 Martha C. White, "So Long, Middle Class: Middle-Income Jobs Are Disappearing the Fastest," *NBC*, August 5, 2016.

322 C.B. Frey and M.A. Osborne, "The Future of Employment: How Susceptible are Jobs to Computerization," Oxford Martin School, September 17, 2013.

323 World Inequality Database, USA.

324 Martin Gilens and Benjamin Page, "Testing Theories of American Politics: Elites, Interest Groups, and Average Citizens," *Perspectives on Politics* 12, No. 3. (September 2014): 564–581.

325 Fredrick E. Allen, "Super-Rich Hide $21 Trillion Offshore, Study Says," *Forbes*, July 23, 2012.

326 "Amazon Inc. Paid Zero in Federal Taxes in 2017, Gets $789 Million Windfall from New Tax Laws," *Institute of Taxation and Economic Policy*, February 2018.

327 Bob Bryan, "We Know More Details About the Final GOP Tax Bill—Thanks to a Lobbyist Who Sent It to a Top Democratic Senator," *Business Insider*, December 1, 2017.

328 "Briefing Book," Tax Policy Center, June 1, 2017.

329 Ben Berkowitz, "Warren Buffet Calls for a Minimum Tax on the Wealthy," *Reuters*, November 26, 2012.

330 Alicia Parlapiano, "How the 'Small-Business Tax Cut' would also be a Tax Cut for the Wealthy," *The New York Times*, December 20, 2017.

331 David Squires and Chloe Anderson, "U.S. Health Care from a Global Perspectives: Spending, Use of Services, Prices, and Health in 13 Countries, The Commonwealth Fund," *The Commonwealth Fund*, October 8, 2015.

332 Adam Gaffney et al., "Moving Forward from the Affordable Care Act to a Single-Payer System," *American Journal of Public Health* 106, No. 6 (June 1, 2016): 987–988.

333 Jason Kane, "Health Costs: How the U.S Compares with Other Countries," *PBS*, October 22, 2012.

334 Lawrence Mishel, Elise Gould, and Josh Bivens, "Wage Stagnation in Nine Charts," *Economic Policy Institute*, January 6, 2015.

335 Ashlea Ebeling, "Final Tax Bill Includes Huge Estate Tax Win for The Rich: The $22.4 Million Exemption," *Forbes*, December 21, 2017.

336 Gene Sperling, "Don't Cut the Estate Tax—Raise It," *The Atlantic*, April 25, 2017.

337 Ingraham, Christopher, "Somebody Just Put a Price Tag on the 2016 Election," *Washington Post*, April 14, 2017.

338 Ken Jacobs, Ian Perry, and Jenifer Marcgillvary, "The High Public Cost of Low Wages: Poverty-Level Wages Costs U.S. Taxpayers $152.8 Billion Each Year in Public Support for Working Families," University of California, Berkeley, Labor Center, April 2015.

339 "A Shocking Student Loan Debt Statistics for 2021," Student Loan Hero, January 2021.

340 Hannah Fern, "Now Canada Is Trying Basic Income, Britain Can Ignore It No Longer," *The Independent*, March 8, 2016.

341 Jason Hickel, "Basic Income Isn't Just a Nice Idea. It's a Birthright," *The Guardian*, March 4, 2017.

342 Lucia Graves, "Their Own Media Megaphone: What Do the Koch Brothers Want from Time," *The Guardian*, November 27, 2017.

Chapter 8: Overpopulation: The Elephant in the Room

343 The Statistics Portal, "Population Projections of the United States from 2015 to 2060 (in Millions)," https://www.statista.com/statistics/183481/united-states-population-projection.

344 https://www.indexmundi.com/g/r.aspx?v=24.

345 Pew Research Center, "The Future of World Religions: Population Growth Projections."

346 Rick Leblanc, "The Environmental Impacts of Overpopulation," *The Balance*, August 31, 2018.

347 www.populationconnection.org.

348 Heather Alberto, "Debunking Overpopulation," *Ecologist*, April 16, 2020.

349 Worldometers, "Regions in the World By Population," www.worldometers.info/world-population/population-by-region/.

350 *Ibid.*

351 Religious Landscape Study, Pew Research Center

352 Michael Lipka and Conrad Hackett, "Why Muslims Are the Fastest Growing Religious Group," Pew Research Center, April 16, 2017.

353 United Nations, Department of Economic and Social Affairs, Population Division (2015), "World Fertility Patterns 2015," Data Booklet (ST/ESA/SER.A/370).

354 Farzaneh Roudi-Fahini et al., "Women's Need for Family Planning in Arab Communities," Arab States Regional Office, July 2012.

355 "Muslim Identity and Ethnicity Hindering a Rural Fertility Transition," International Population Conference, Marrakech, Morocco, October 2009.

356 Russell Goldman, *New York Times*, May 17, 2017.

357 "Estimated Population Growth in India from 2010 to 2050, By Religion," Statista Research Department, October 6, 2020.

Appendix: Fossil Fuels

358 http://environment.nationalgeographic.com/environment/global-warming/high-cost-coal/.

359 http://www.eia.gov/countries/country-data.cfm?fips=ch.

360 http://data.worldbank.org/indicator/EG.ELC.COAL.KH.

361 http://www.resourcesandenergy.nsw.gov.au/__data/assets/pdf_file/0009/182484/International-Mining-Fatality-Database-project-report.pdf.

362 http://www.cdc.gov/niosh/nas/rdrp/appendices/chapter4/a4-55.pdf.

363 "Black lung on the rise among US coal miners," *World Socialist*, January 11, 2010.

364 Union of Concerned Scientists, http://www.ucsusa.org/clean_energy/our-energy-choices/coal-and-other-fossil-fuels/how-coal-works.html#.VTLW3yHBzGc.

365 U. S. Environmental Protection Agency, 1997, "Mercury Study Report to Congress," Volume II: An inventory of anthropogenic mercury emissions in the United States.

366 http://www.scientificamerican.com/article/coal-ash-is-more-radioactive-than-nuclear-waste/.

367 http://www.usatoday.com/story/money/business/2014/06/08/coal-plant-retirements-barely-cut-carbon-emissions/10008553/.

368 http://www.ucsusa.org/global_warming#.VVi6lvlViko.

369 http://unfccc.int/ghg_data/items/3825.php.

370 http://www.scientificamerican.com/article/bpa-may-prompt-more-fat-in-the-human-body/?WT.mc_id-SA_WR_20150603.

371 Mason Inman, "Natural Gas: The fracking fallacy," *Nature*, Volume 516, Issue 7529, December 03, 2014.

372 http://www.nrdc.org/energy/gasdrilling/.

373 Mark Peterson et al, "Incorporating Induced Seismicity in the 2014, United Seismic Hazard Model–Results of 2014 Workshop and Sensitivity Studies," USGS Open Fie Report 2015-1070.